Transitioning Island Nations Into Sustainable Energy Hubs:

Emerging Research and Opportunities

Catalina Spataru
UCL Energy Institute, UK

Published in the United States of America by
IGI Global
Engineering Science Reference (an imprint of IGI Global)
701 E. Chocolate Avenue
Hershey PA, USA 17033
Tel: 717-533-8845
Fax: 717-533-8661
E-mail: cust@igi-global.com
Web site: http://www.igi-global.com

Library of Congress Cataloging-in-Publication Data

Names: Spataru, Catalina, author.
Title: Transitioning island nations into sustainable energy hubs : emerging research and opportunities / by Catalina Spataru.
Description: Hershey, PA : Engineering Science Reference, [2019] | Includes bibliographical references.
Identifiers: LCCN 2017060579| ISBN 9781522560029 (h/c) | ISBN 9781522560036 (eISBN)
Subjects: LCSH: Islands--Energy conservation. | Energy policy--Developing countries. | Renewable energy sources--Technological innovations. | Energy development.
Classification: LCC TJ807.9.I75 S63 2019 | DDC 333.793/209142--dc23 LC record available at https://lccn.loc.gov/2017060579

British Cataloguing in Publication Data
A Cataloguing in Publication record for this book is available from the British Library.

Table of Contents

Preface

Islands nations are among the most vulnerable to climate change. Islands are facing challenging times and some major policy dilemmas. According to the Amsterdam Treaty, declaration No. 30, "... insular regions suffer from structural handicaps linked to their island status, the performance of which impairs their economic and social development". In this context, it is important to note that even today many of the islands have considerable energy supply problems and relies on oil-based power generation. On the other hand, both diversity between different island regions and their worldwide distribution, makes them excellent demonstration sites and vehicle for diffusion of innovative energy solutions that can effectively replace the conventional methods. Article 174 of the Lisbon Treaty recognizes that island regions as a whole face practical handicaps that require special attention. Nevertheless, the European Commission (political declaration on clean energy for EU islands May 2017 (European Commission, 2017) considers there is a need for further work to secure the adoption of an appropriate strategy for addressing the specific needs of islands. "Islands are well placed to employ innovative solutions and attract energy investments that integrate local renewable production, storage facilities and demand response".

The concept of "sustainable development" was initially introduced in the Brundtland Commission Report in Our Common Future (Hinrichson, 1987) as follows: "A process of change in which the exploitation of resources, the direction of investment, the orientation of technological development, and institutional change are all in harmony and enhance both current and future potential to meet human needs and aspiration."

Encouraging increase of renewable energy technologies is seen as the key driver to achieving a wide range of sustainable goals in regional islands. While large number of seasonal holiday makers put pressure on both local cultures and environments, regional and national policies often do not support small scale development.

Within this content, the objectives are: to help islands nations seeking to become sustainable energy nations; help establish innovative ways for integrating renewable energies, with adequate business models, and provide understanding and awareness about current experiences, potential and advantages of renewable energy utilization and energy efficiency in the islands nations.

In this book, the author shows that islands represent great opportunity for demonstrating energy solutions, offering various advantages as laboratory for studying renewable energy sources, can be a great source for export to mainland or other neighboring islands and countries. It provides an understanding of current status in islands worldwide and potential future solutions with several examples assessing the technical and economic potential. It contains several case studies. It creates an inventory of existing information and recent research, and compiling a comprehensive guide, providing a baseline understanding of developments and challenges on islands.

The book will be an essential reading for policymakers, investors, professionals, energy systems operators and planners, engineers, researchers, anyone concerned with energy and security internationally and with integration of renewables, with islands as case studies.

It will discuss the prospect of different islands as an energy hub for new technologies, from wave and tidal power technologies to wind and solar. Technological, geographical and commercial factors will be taken into account when discussing potential change of islands into versatile 'energy hubs' for immediate and longer-term renewables operations. Successful development of the energy hub will bring valuable longer-term skilled employed opportunities and benefits not only to those communities but also at national level.

There are already several programmes and policies launched towards the direction of applying different technologies to the islands. Various islands projects are in progress or have already been implemented worldwide. Therefore, there is a need of understanding of existing work in this area towards a better coordination with strategic frameworks to answer new research questions. This book aims to provide a deep understanding of islands and develop insights into the type of changes needs to be made: techno-economic, policy, market and political changes to implement renewables in islands towards a sustainable future.

Islands have locally most of the infrastructures needed for the management of their resources. However, during the season, because of increase in tourist number, islands takes a heavy toll on both infrastructures and resources. There is growing evidence that islands can function as laboratories of technological, social and financial innovation, as test-beds for the deployment of innovative integrated solutions. Such solutions could be use low carbon technologies, increase use of renewable energies, use of energy storage technologies, information communication technology (ICT) and so on, solutions which have the potential to maximize the synergies between energy, transport, water and waste management. Testing such solutions, could then be implemented in other generally geographically isolated areas of the mainland such as small municipalities, mountainous, rural.

The aim of this book is to give a comprehensive understanding of current energy situation in islands and develop insights into the type of changes needs to be made to make islands the "energy hubs" of the future, the "hottest energy centred points", the "golden spots" for sustainability. In addition, the book means to be helpful wherever in the world, since there is a general interest in understanding the environmental change.

Practitioners, engineers, researchers, politicians, policy makers could be interested in learning about the diversity and possibility of using islands as future energy hubs. For students at undergraduate, MSc and MRes level will provide an understanding of methods, concepts and practice of renewable integration, offering a combination of theory, concepts and practice, with innovative practical case studies, and understanding how to model an energy system.

ORGANIZATION OF THE BOOK

The book is organized as follow:

"Introduction" sets the scene on the topic of the book. To provide a foundation, Chapter 1 ("Islands Diagnosis") provides a general overview of the environmental concerns, challenges and opportunities for clean energy transformation islands. This is followed by a detailed review of the initiatives worldwide regarding environmental, energy, threats (e.g., natural hazards), economical, climate change, general development problems in Islands is covered in the chapter. This helps the reader to have an overview of the state-of the art on different issues associated with Islands. Finally, the topic related to policy and strategic planning initiatives worldwide are presented.

Chapter 2 ("Interconnections and Islands: Global Experiences and Learning") covers a historical perspective of islands interconnections, submarine power cables interconnections and technologies, projects, finishing with a global mapping of islands interconnections. This helps the reader to understand historical and current progress, trends, challenges and potential for innovative solutions.

Chapter 3 ("A Global Perspective on Experiences and Practices for Low Carbon Technologies and Renewable Energy in Islands") describe practices from all over the world in terms of renewable energy in islands, islands test beds and islands in transitional phase. In essence, this chapter provides to the reader an understanding on how some islands have managed to successfully accomplish large proportions of integrating renewable energies, which in turn may help them, for example, to attract tourism and improve the local economy and the quality of life of its inhabitants.

Chapter 4 ("Business Models and Policy Measures to Support Renewable Energy Community in Islands") introduces an overview of some basic concepts of Business Models (BMs), and discussions on different BMs for energy applications on Islands. Also, some risks associated with the implementation models are summarized and some discussions provided.

Furthermore, since islands are considered paradises for tourism, I have also discusses key concepts and present some case studies of sustainable tourism.

Chapter 5 ("Indicators, Modelling, and Visualization of Islands") provides an understanding on indicators, modeling and visualization of islands ' results. It starts with a review of existing modeling approaches for islands, and continues with a description of a 'toy' model for sustainable islands called S-Isla, the application of the model to a case study (Crete).

Chapter 6 ("Islands: Sustainable Hubs of the Future") discusses the possibility of transitioning islands as sustainable hubs of the future. This chapter provides a discussion on the topics provided in the previous chapters and it introduces the concept of energy hub and the potential of using it for islands.

The essential dependencies between chapters indicate the importance of the subject, highlight and discuss specific challenges and potential solutions. Learning from the past and understanding the present we can plan better for the future. This book enables the full understanding of challenges in islands and across sectors, so that the impact of the interaction in different areas on each other can be fully tested.

REFERENCES

European Commission. (2017). *Political Declaration on Clean Energy For EU Islands*. Retrieved from https://ec.europa.eu/energy/sites/ener/files/documents/170505_political_declaration_on_clean_energy_for_eu_islands-_final_version_16_05_20171.pdf

Hinrichsen, D. (1987). *Our Common Future: A Reader's Guide The Brundtland Report Explained*. London: Earthscan Pub. Retrieved from http://www.are.admin.ch/are/en/nachhaltig/international_uno/unterseite02330

Acknowledgment

Perhaps you wonder why a book about islands is written by a single author. I can ensure you that is not the case and that there were many who helped me see this book through to completion. My first supporters were my passion for islands and this idea I had for years, which pursue in my head until the completion of the book.

Secondly, I want to thank to all my students (unfortunately too many to be able to mention all of them here) for taking my modules and participating to lectures, to discussions and express their view. Special thanks goes to my researchers, Eleni who collected some of the historical information, to Tola and Selman who help with some graphs.

I was fortunate enough to have two excellent manuscript reviewers who pointed out problems, identified difficulties and provided a constructive feedback. Thank you!

Special thanks goes to my editor Marianne Caesar being supportive and understandable during the duration of the project.

I thank also to my life helpers Andrei and Anya, my parents, my friends and colleagues. Their friendship, support, and encouragement are most meaningful.

Introduction

I have a great passion for islands since I was born. This book was motivated by the desire I have had to provide a framework for learning to my students. The provocative challenge is to make students understand the use of mathematical models, to describe the reality and provide quantitative and qualitative answers to questions raise by practitioners and policymakers. My challenge over the years has been to show students, researchers and practitioners how well the theories that guide our thinking, acknowledged or not, fit the facts and how models using the theory describe the reality. This book is the result of many years of facilitating, researching and teaching energy systems. Having these in mind, part of my modules *Smart Energy Systems* and *Metrics, Modelling and Visualisation of the Resource Nexus*, I provide students with the conceptual foundation and introduce them to the skills necessary for understanding energy systems and the resource nexus (water-energy-land-food-materials) on islands and to verify them on case studies. So far together with my students and my researchers, we analysed more than 200 islands.

When we look at the islands, we see the beauty of them and as paradise places. In reality, they suffer various challenges. The central argument of this book is that islands are opportunities for demonstrating energy solutions and could provide a great source of information and solutions, which can be also adopted by other geographically isolated areas of the mainland (small municipalities, villages). Furthermore, they have capacity in becoming energy hubs of the future, supporting through interconnections with the mainland the neighbouring countries. When the interconnection with the neighbouring countries is difficult, they can also become self sufficient energy hubs of the future. Despite the many challenges they face, there is a need of understanding existing work in this area, and better coordination with strategic framework to answer new research questions.

Introduction

The title of this book speaks to two fundamental current research and opportunities and discusses emerging research and opportunities regarding the sustainable development of islands that are nowadays confronted with a number of severe challenges. It contains a number of features to enhance reader learning: extensive use of examples and a wide range of information compiled to provide an understanding of opportunities for islands. Furthermore, it depicts steps of such a transitioning by current practices, capitalizing on these steps in order research aspects as well as opportunities to be sketched.

The objective is to provide a book which could reach potential audience, such as those conducting research on different topics associated with environmental concerns and tourism on Islands. It could be of immense value to those involved different areas of research associated with energy supply, tourism, environmental impact, climate change, business models, etc. It also has the potentiality to assist, for example researchers and postgraduate students, practitioners, environmentalists, and decision makers in those islands interested in developing their economy based on tourism, which may help to enhancing their quality of life. Moreover, the book aims to be useful everywhere in the world, since there is a universal interest on sustainability and islands. Furthermore islands could be used as energy hubs producers for mainland to reach sustainable development goals. Literally, there are thousands of islands worldwide facing environmental problems and eager to attract tourism to enhance their quality of life. At the same time, this attracts the attention to scholars from different background in conducting research on the subject.

The companion website (www.islandslab.com) has additional information on islands and provides energy resource snapshots of islands. The website is regularly updated with new features and content. I have also created the Smart Global Islands Forum group on LinkedIn which promotes sustainable solutions and development in islands nations globally.

Chapter 1
Islands Diagnosis

ABSTRACT

In this chapter, the author provides an overview of the great challenges that islands are facing. It consists of an introductory part, raising environmental, economic, social, climate change, biodiversity, issues, etc. revealing the high risks to which island regions are exposed. It then discusses the way renewable energy exploitation can be a promising option for reaching sustainability objectives in general and coping with the above challenges in particular. This chapter exploits the information on peculiar challenges faced by islands nations with respect to sustainability, the proposed plans, challenges, and the issues raised by these challenges, the reaction in terms of projects and policy initiatives to cope with these challenges, and identification of sectors for paving a sustainable future of island nations. Despite the fact that numerous islands have presented and introduced target plans to advance renewable energy deployment, the policy design and execution is frequently lagging. Furthermore, there are still significant remaining things to be done.

BACKGROUND

The origin of the word 'island' derives from a prehistoric Germanic aujō which denoted 'land associated with water', and was distantly related to Latin Aqua 'water'. This goes to Old English īeg island, which together with 'land' form 'īegland'. By the late Middle English period this had developed to iland, the form which was turned into island (Ayto, 1990).

DOI: 10.4018/978-1-5225-6002-9.ch001

There is no clear official definition for what constitutes an island nation. An island is characterized as a land mass totally encompassed by water. Land masses which are surrounded by water are refer to archipelago, key, reef, or rock. Islands are found in Oceans, Seas, Lakes and Rivers everywhere throughout the world.

There are different types of islands, in various shapes, of all sized and geographical attributes. Beate (2018) provided a comprehensive review and understanding of geography of small islands. Charles Darwin in his famous book "On the Origin of Species" (1859) was the first one who tried to categorise islands. Later in 1880, Alfred R. Wallace came up with a categorization of islands based on their geographical structure, differentiating between continental islands and ocean islands (Wallace, 2012). Since then many scientists tried to classify islands based on their genesis, but only few classification are based on the fundamentals of island formation such as: character of the underlying crust (Grigor'yev 2014). The geographer Christian Depraetere (2008) listed 86,732 islands with a size of more than 0.1 km^2 and for islands below the size of 0.1 km^2, the author applied fractal models, estimating an additional 700 million of islands to exist globally (Depraetere & Dahl, 2007). There are 193 sovereign states officially recognized by the UN (2016). From this 47 are island states, the largest one being Indonesia.

Some broad attributes of an Island comprise in unique species of animals and exceptional vegetation. For instance, Madagascar has a unique group of primates called lemurs. Giant Tortoises of Aldabra can be found in the Galapagos Islands, while the Komodo Dragon is found just on Komodo Island. Some are endemic which implies they are not to discover anyplace else. They don't migrate. Such illustration is the white starling of Bali. Regarding vegetation, there are approximately 560 Species of Galapagos Plants living in the Islands, from which a third are one of a kind to the Islands. These include: giant prickly pear cactus, cheery-like tomato, an individual from the daisy family that privately advanced into 17 distinct Species (genus Scalesia) and so on. On Catalina Island, Catalina endemic Plants are Species that happen normally on the Island and no place else in the World.

A group of scientists at the Australian National University studying tiny grains of minerals say that 4.4 billion years ago, Earth was a barren, mountainless place, and practically almost everything was under water (Burnham & Berry, 2017). In the previous couple of hundreds of years, small islands were viewed as segregated ranges, in some cases served as military reservation, training purpose, for prisoners, immigration station and so forth. There is captivating hidden history for many of these islands. These days, islands are seen mostly as holiday paradises, being of high interest for worldwide tourism. Taking a look at the current trends in the tourism industry, islands are becoming increasingly popular, with TUI customers now spend their holidays on Canary Islands, Cape Verde Islands, Maldives and so on (TUI Holiday Atlas 2016). For quite a long time, people living in islands cleverly harnessing their own resources to survive. A large proportion of these resources come from nature. They have dependably needed to create ingenious ways of harnessing the sun, wind, the biomass and the water available at their disposal. Looking back at the history, we can't overlook how small waterfalls were ingeniously harnessed on islands as Corsica, Yakushima and so on; how wind power helped pumping seawater in places such as, Ibiza, Lanzarote, Djerba and so on changing them in wealthy centres of trade. Despite the ingeniously harnessing of natural resources, a significant number of island nations face significant challenges because of environmental change. These are discussed in the next section.

CHALLENGES DUE OF THE ENVIRONMENTAL CHANGE

It was determined that 2016 averaged temperatures were with 0.99 degrees Celsius (1.78 degrees Fahrenheit) warmer than the mid-twentieth century mean (NASA, NOAA Data). With an estimated of 95 percent certaintly, 2016 was determined to be the hottest year (NASA estimates, 2016). Phenomena such as El Niño or La Niña, which warm or cool the upper tropical Pacific Ocean and cause variant in atmospheric condition patterns, contribute to short-term varieties in worldwide normal temperature. A warming El Niño event was in effect for majority of 2015 and the main third of 2016. Specialists (Niedzielski, 2014) appraise the immediate effect of the El Niño. The tropical Pacific extended the annually worldwide temperature irregularity for 2016 by 0.2 degrees Fahrenheit (0.12 degrees Celsius).

One of the central concerns with respect to environmental change is global sea-level rise and its effect on islands. It has been already seen the effect of sea-level rise on coastal land loss on many islands (Forbes et al., 2013). They confront dangers because of ocean levels rise, pulverizing tropical storms, hurricanes and so on. NOAA (NOAA, 2016) estimates a rate of rise of 0.12 inches per year since 1992. Between 1992 and 2010, worldwide mean sea level rose, with the most noteworthy rise in the Western Pacific.

ESA CCI project, who produced a territorial outline map of mean sea level patterns from October 1992 to March 2010 as measured by radiolocation altimeters, shows a few regions encounter a rise of up to 10 mm per year, while few others see a decrease of about 10 mm per year. On average, the global sea level rises almost 3 mm each year. The sea levels are specifically caused by the development of water as it warms and the melting of polar ice caps (National Aeronautics and Space Administration NASA, 2016). Because of those, islands go through transformation in terms of shape.

A few islands are at chance of losing seashore arable land. As quickly as the restrained accessible soil on those islands is salinified it seems to be exceptionally difficult to create subsistence plants, for instance breadfruit. Additionally local fisheries are influenced by higher sea temperatures and increased ocean acidification. As sea temperatures rise and the pH of oceans will increase, many fish and different marine species would die out or change their haits and variety. Warming of oceanic water masses, leads to changing patterns of fish stock. Modifications in atmospheric temperatures and precipitation impact the composition of vegetation and fauna. Water supplies and local ecosystems, for instance, mangroves, are undermined by a dangerous atmospheric deviation (GIZ, 2016).

Due to increased events of extreme weather (hurricanes and droughts), the tourism sector has suffered drastically. Hurricanes and cyclones activity appears, by all accounts, to be expanding. In 2015 and 2016 two category-five cyclones hit the Pacific: Cyclone Pam hit Vanuatu (March 2015) and Cyclone Winston hit Fiji (February 2016). The latest one, Winston was determined as the strongest tropical cyclone to hit the Southern Hemisphere in recorded time. In 2017 after Hurricane Irma crushed the Caribbean, another to a great degree risky class five storm threated the region. Hurricane Maria – the fourth named storm since Irma – has just caused extensive damage across Dominica, towards Puerto Rico and the British Virgin Islands. The storm was the strongest in almost a century. Just few days after the Hurricane Irma which strongly hit the area, Hurricane Maria hit the area too, which positioned as the most intense Atlantic storms on record. This produced significant destruction on several Caribbean Islands. At risk are especially low lying islands with coastal population centres. A considerable lot of the Pacific Island nations are confronting the issue of rising sea levels. Such a case is The Maldives an archipelago of islands and atolls located in the Indian Ocean. The local economy is significantly undermined by environmental change. The tourism sector is being threatened by the improved likelihood of violent storms, harm to coral reefs, and beach erosion. Just a couple of choices are accessible:

islanders can either leave the areas, move to higher ground, or build walls to hold back the sea. The Maldives have officially executed a few measures to combat sea level rise including: repairing local infrastructure, especially ports and building a system of protective walls around the capital of Malé at a cost of US$4,000 per meter, financed to a great extend by Japan (UN Country Team in Maldives, 2005). These protective walls manage to save the capital from the great tsunami that struck the Indian Ocean in December 2004. Islands exposed to higher wave energy in addition to sea level rise experienced particularly loss compared with more protected islands.

The scientific community points out t4hat the sea level rise, which has accelerated since the mid 1990s will continue throughout the following century. The worst case scenario predicts an average increase in the sea level of +45 to +82 cm between now and 2100 (Church et al., 2013). Moreover this pattern is irreversible halfway as a result of the latency phenomena that characterise the oceanic and atmospheric processes. Some of Kiribati's islands may disappear under the sea this century. Tuvalu, a West Pacific nation whose peak height rises only 5 meters over sea level is and no more in risk could be unpleasant inside 50 years. Situated in the middle of Australia and Hawaii, encompassed by an expansive number of miles of vast sea, Tuvalu includes three reef islands and six atolls in the South Pacific Ocean. In 1972, Cyclone Bene wiped out essential vegetation and tree trims through saltwater saturation of fertile soil. Additional stress on Tuvaluan food production has been seen due to increased temperature and acidification of oceans. Acid weaken local feeding reefs and the reinforcement of shellfish, whilst heat bleaches the coral and reduces survival rates among heat-sensitive species. Kiribati's chain of 33 atolls and islands could become uninhabitable as early as 2030 due to raise of see level. A comparable destiny could likewise fate the Maldives. The destiny of small island states might be seen as a microcosm for what environmental change might mean for whatever is left to us at its generally extreme. Island states fights hard for the optimistic 1.5 degrees Celsius limit on emissions rises at the global climate change conference COP21. Despite the fact that postcards shows delineate tranquil peacefulness places, behind these nations' enticing facades lie varied challenges. An anticipated global mean sea-level rise of one metre or more until the end of the century would require effective adaptation within the vulnerable areas- coastal zones and small islands (Nurse et al, 2014). Without substantial and sustained reductions of greenhouse gas emissions, worldwide we will experience more extreme weather events and sometimes with high consequences. Adaptation is essential, local and regional communication and cooperation are vital to build resilience even with challenges from an evolving atmosphere. Large-scale sea defences, are barely implementable in isolated and small island regions. Financial and physical resources are inaccessible and the basic system is unsustainable since expansive large structural sea defences often times evoke extra problems (e.g. erosion), which might got to be managed afterward on. Environmental change analysis has fundamentally centered on understanding causes and illustrating future circumstances. Impacts have continuously been on the motivation but have been related with high degrees of vulnerability. Detailed information and data about the impacts of environmental change are not really accessible for small islands. In addition to environment, there are also peculiar challenges faced by island nations with respect to sustainability objectives, which are discussed in the next section.

PECULIAR CHALLENGES FACED BY ISLANDS NATIONS WITH RESPECT TO SUSTAINABILITY OBJECTIVES

Heavy dependence on volatile export markets, remoteness, dependence on fossil fuels are few of the challenges faced by islands nations. Those islands which are small, have limited options for economic growth and market extension. For them to pursue sustainable development depends on natural resources and the way the harvest them, income and finance (Bass & Dalal-Clayton, 1995).

Islands' power systems are often isolated, with no interconnection to the mainland or to other islands. They face a number of limitations in terms of market liberalization and unbundling which usually fail to become implemented. As a consequence, new investments are usually limited. Islands possess fragile ecosystems while facing challenges related to un-sustainable resource management including water and land. Massive waves of tourism, especially on certain regions (for example in the Mediterranean basin, in the Caribbean islands or in the Pacific islands) cause significant discrepancies in energy demand on a monthly basis while placing risks in the smooth operation of the local grid (Kaldellis et al., 2012). However, each island depending on their size, economic structure, population and location is unique, encountering different barriers and drivers towards sustainable development and requires tailor-made policies and regulations to drive transition towards sustainable energy and resource management.

Their main power source for non-interconnected islands and autonomous power systems is usually diesel and heavy fuel oil. While this is usual for isolated systems, it is neither sustainable, nor enhances security of supply in those regions. Preserving isolated power systems is a fourfold challenge for islands related with: power demand discrepancies between summer and winter causing frequent power cuts, fragile interconnections, fuel dependency, and low levels or renewable energy sources integration. Power generation costs, are driven by oil prices with significant fluctuations, while price speculations cannot be forecasted, resulting in island operators' financial inflexibility to adopt sustainable options.

For example the electrical energy prices for end-users within insular areas in Pacific varied between 25-34 cents per kWh comparing to the mainland United States where the same cost ranges between 10 – 14 cents per kWh in 1982 (Pacific Power Association, 2006). It has been also seen significant volatility in price in a short period of time. Such case is the American Samoa island, where the price for electricity for end-users vary from 31 cents per KWh in September 2010 to 40 cents per kWh in December the same year. The volatility in price was due to higher and variable cost of fuel per barrel due to transportation cost, increase cost of maintenance due to aging of electric production facilities (Pacific Power Association, 2006). Some countries support the difference in prices between insular areas and mainland through public funds to ensure the same costs. This is enhanced by barriers from the market monopoly and the lack of economies of scales. Furthermore, the internal electricity market particularly in small islands suffers from regulatory gaps since often operates under the same regulations with the National Grid System (NGS). This is enhanced by lack of economies of scale and barriers stemming from the market monopoly which usually exists (Eurelectric, 2012). On top of that, the smallest islands have to conquer the shortage of fresh water completing a series of structural handicaps (Marin et al., 2005; Kaldellis et al., 2012) .

In addition, in terms of environmental sustainability, oil, having a carbon intensity equal to 73.3kgCO_2/GJ (Blanco et al., 2014) leads to emission levels higher than the approved ones (Kasselouri et al., 2011). This will impact on health and quality of life. Therefore, both economic and environmental impacts of fossil fuels use in islands must be examined in detail to provide a great understanding of where changes should be made, what kind of changes and under which conditions and compromises.

OPPORTUNITIES AND CHALLENGES FOR CLEAN ENERGY TRANSFORMATION

Islands are open entryway for illustrating energy solutions. Islands' characteristics give a few preferences as research facilities for examining 100% renewable energy sources systems. They might represent an extraordinary source for export to mainland or other neighbouring nations.

In Europe for instance, there are 2,425 of inhabited islands (small, big, bridged, un-bridged, NUTS, non-NUTS, in seas, states, regions, municipalities or other) with a resident population of 13,9 million people (Baric et al., 2016). Their power systems are frequently isolated, with no interconnection to the EU mainland or to different islands. Their business sectors are constrained, that makes investment difficult to justify. Besides, their natural circumstance is delicate and additionally weakened by non-sustainable tourism, notably on some Mediterranean islands. The regulation varies from one island to another, contingent upon the particular connection between the pertinent member states and the EU. Greece has the biggest number of islands, 6.000 of which only 227 are inhabited. From these, 55 are non-interconnected islands (Zafeiratou & Spataru, 2017). From 227 inhabited islands, 130 are small islands with a total of 65,552 inhabitants (Baric et al., 2016). The electricity generation for these islands consists of 32 autonomous power stations (APS) using diesel, light or heavy fuel oil as their primary energy source. A few issues emerge because of the broad utilization of such fuels. The most critical issue is the high concentration of emissions pollutants. The non interconnected part, more than half of the total national thermal plants, have emissions higher than the approved ones and shares responsibility regarding the way that Greece will most likely not achieve the requirements for emissions and energy reduction for 2020.

A few projects and arrangements have been launched towards the direction of applying distinctive innovations to the islands. Different island projects are in progress or have already been implemented in Europe. Such illustrations are: Samsoe island which has been picked up by the Danish government as a 100% renewable energy island (Hermansen); Madeira island (Portugal) which has high levels of renewable generation, promoting sustainable tourism; Canary islands in Spain increase their wind and solar capacity as a whole.

There is a further need of understanding of existing work in this area, and better coordination with strategic frameworks to answer new research questions.

The following challenges have been identified:

1. **Capacity- Building Design of Decarbonisation Plans:** While the broad similarities have been well documented, the situation changes drastically as you narrow the spatial scope. For example the same transition pathways that are implemented in the islands in the Mediterranean Sea may not work for islands located in Scotland. Furthermore, the optimal mix of policy and operational changes will depend on local conditions and should be tailored to local needs and institutions to be effective in practice. A solution to these challenges is a catalogue of current energy activities and institutional relationships, including policies and regulations, incentive programs, bilateral and multilateral development partners; bodies responsible for permitting new energy projects. Such catalogue will be a useful reference and will help to understand future opportunity pathways.

To define integrated decarbonisation plans, it requires to establish an energy baseline by sector and time of year including information of the current energy system production levels and costs including energy production, line losses, common end-uses behavior and peak loads. Compiling and analyzing this information depends on access to information. Having accessible and accurate information on current energy system, we will be able to determine where changes should be made and quantify them.

2. **Cooperation With Stakeholders and Access to Data and Information:** One of the biggest challenges is cooperation with stakeholders and access to data and information to understand to understand problems and failures in projects implementation; understand legal risk stems primarily from changes in law, or the application of law, that would negatively impact the project. Gaining access to internal documents is challenging. Apart from being time-consuming, the collection of the data may proof difficult due to being protected and unavailable to the public.

The problem of the availability of data could be tackle by carrying out a survey. Furthermore, an island database, together with an efficient geographic information system can help to measure and quantify the changes that need to be made and to provide the possibility for a more accurate future planning.

Furthermore, selecting cases per cluster group could demonstrate the existence of strong approach which could be easily replicable.

There are also a number of challenges related to identification of individual projects (incl. technical, economical, political feasibility): such as changes to the schedule, budget, staff availability, changes in law and its application that would negatively impact the project, performance, cooperation and payment risk.

Some of these challenges can be addressed. For example, the cooperation problems can be reduced through meetings and training. Information on causes of specific risks need to be compiled and assign them a probability value to give a clear indication of what doesn't work according to the plans.

A few projects and arrangements have been launched towards the direction of applying distinctive innovations to the islands. Different island projects are in progress or have already been implemented in Europe. Such illustrations are: Samsoe island which has been picked up by the Danish government as a 100% renewable energy island (Hermansen); Madeira island (Portugal) which has high levels of renewable generation, promoting sustainable tourism; Canary islands in Spain increase their wind and solar capacity as a whole. There is a further need of understanding of existing work in this area, and better coordination with strategic frameworks to answer new research questions.

3. **Features Changes:** Trends in rate of population growth, the level of facilities and infrastructure, geomorphologic and natural conditions of islands, and behaviour of many endogenous factors should be taken into account when implementing support policies.
4. **Limits of Development:** Increase in population, entails impact on environment due to high demand.

This imposes a genuine limit on the economic and social development of islands.

One of the most appropriate methods in collecting appropriate and necessary information is mapping and use of geographical information systems.

5. **Reliance on Tourism:** Tourism is considered as a vital source of income for the majority of the islands in the world (Manning, 2016). Island complexes such as the Greek islands, the Balearic Islands, the Hawaiian Islands, the Galapagos Islands, the Canary Islands, the French Polynesian Islands, and the Caribbean islands rely on tourism as the principal sector in the economy (Sheldon, 2005). In half of the SIDS,

> tourism expenditure accounts for over 40% of all their exports of goods and services. Moreover, it is linked with more than 20% of GDP in 40% of the of SIDS according to the data availability. The importance of tourism in island economies is demonstrated by the recent graduation of Cape Verde and the Maldives from Least Developed Country status due to their levels of income from tourism (UNWTO, 2013). Tourism is also economically significant since it is a source of foreign trade. A number of the SIDS would announce generally expansive adjust of instalments deficits in need of the assess and tax declarations from tourism. There are moreover a few focal points which are not straightforwardly financial, but which have an affect on the quantifiable well-being of the local population (Briguglio & Brigulio, 1996). Tourism has a key role to the national balance of payments and the creation of jobs on island, in any case it misuse the environmental and energy resources as well as having impacts on the neighbourhood culture (Manning, 2016). Consequently, counting the 52 Small Island Developing States as well as thousands of islands around the world, special emphasis must be placed on the tourism sector and the potential it has to stimulate the green economy in those regions. This is further enhanced by the global trends showing that tourism is expected to grow rapidly in the coming year overtaking economic growth factors (Manning, 2016; UNWTO, 2013). In particular, projections show that the international tourism arrivals are forecasted to reach 1.8 billion by 2030 (UNWTO, 2013).

The main concept of sustainable tourism is balancing financial and environmental activities. It is well known that numerous, in the event that not all, financial activities have an effect on the environment and that this has a criticism impact on the economy itself. This is particularly so in the case of tourism which uses the environment as a resource. Tourism depends to a exceptionally expansive degree on a area which is charming and alluring to tourists, and negative environmental impacts caused by tourism itself, may therefore have a direct impact on tourism itself (Briguglio & Brigulio, 1996). Sustainable tourism being part of the wider sustainability concept, may be defined as "tourism which is developed and maintained in such a manner and scale that it remains viable in the long run and does not degrade the environment in which it exists to such an extent that it prohibits the successful development of other activities".

The environment agency of the United Nations (UN) jointly with the World Tourism Organization (UNWTO) formulated the following definition for sustainable tourism: "Tourism that takes full account of its current and future economic, social and environmental impacts, addressing the needs of visitors, the industry, the environment and host communities" (UNEP and UNWTO 2005, p.12).

Tourism although it offers several economic benefits to the islands has a direct impact on the local environment, as it is an intensive energy and water consuming sector. In addition, tourism facilities occupy large land areas on the islands. Stressing the fact that small islands usually have exceptionally fragile eco-systems, rich in biodiversity, which have to preserve, while in parallel the majority of tourists concentrate in the coastal areas, they require the highest attention (Goodwin, 2008). The impact of tourism in natural resources can be summarized in the following points:

- **Land Resources and Land Biological Diversity:** Including vegetation and wildlife are threatened due to large tourism infrastructures and facilities. Strong competition between tourism and other sectors (agricultural, recreation areas etc) in small regions due to lack of land areas to accommodate tourists. The massive exploitation of land use leads to increased land prices, deforestation and loss of biological diversity (UNEP Islands, 1996).
- **Waste Management:** For small island developing States, waste treatment and disposal is a significant challenge which requires support. As islands are small areas with limited physical and artificial infrastructure, waste management configures an additional problem. This is worsened by wastes generated from tourism. It must be highlighted that the marine life and coastal regions are endangered by high volumes of waste disposals generated by the tourist sector. Additionally, pollution from ship-generated wastes is an additional significant pollution source for islands. The following factors have been identified as crucial and require immediate action (UNEP Islands, 1996): weak institutional, legislative and enforcement capacities, lack of regional consensus on suitable criteria for sewage, coastal water standards.
- **Coastal Zone Deprivation:** As Tourism activities are mainly located at the coast line considerable impacts on those regions of small island creating States. Especially, in Mauritius, Seychelles, Malta, Cyprus and other islands in the Caribbean, the massive development advancement of tourism facilities along their coastlines brought about in harming the nearby scene and environment (UNEP Islands, 1996). Furthermore,

the annihilation of profitable mangrove woodlands, working as settling places birds and other creature life as well as their execution as a normal obstacle against infringement to the island. Further activities associated with tourism on the islands such as diving, fishing and boating may enhance the pollution of the natural environment.

- **Energy:** The tourism sector is regarded as one of the most consuming sectors as people usually consume energy out of their local residences for entertainment and relaxation without paying attention to the amount of energy consumed. As islands rely mainly on diesel engines and oil as their primary source of energy for electricity, heat and transport the environmental degradation from the intensive use of energy is evident.
- **Water:** Islands face constantly problems of potable water shortage due to needs in the industry, agricultural and residential sector which are worsen during the months of high tourism levels. This is mainly noticed in in the low-lying atolls that have the problems in accessing surface-water and following proceeding to its storage (UNEP Islands, 1996). Furthermore, this is the case generally for the islands with tropical or even Mediterranean climate which frequently suffer from frequent droughts and continuing water scarcity.
- **Climate Change and Sea Level Rise:** All the previously mentioned risks in parallel with the alone standing climate change can have a critical affect on the islands due to the expanding sea level. The Intergovernmental Panel on Climate Change has affirmed a rise of mean sea levels with a pace of 1.5 millimetres per year (Intergovernmental Oceanographic Commission, 1994). Shorelines corrosion deteriorates the resistance of the shoreline with evitable damages to both natural and built environments. Additionally, all the extreme weather events already taking place in those islands are further exacerbated threatening the appealing weather conditions existing on the majority of the islands which is one of their main attractions.

INITIATIVES FROM ALL OVER THE WORLD TO COPE WITH INSULARITY PROBLEMS

This section provides a review of initiatives which took place over time from all over the world to cope with insularity problems. An inventory of these initiatives, their brief description and what they have accomplished is provided below. The inventory is designed as an evolutionary document aiming to explore where the focus is given by the various islands' examples, identify sectors of intervention (energy, tourism, sustainable development, interconnections, renewable energy systems integration and so on).

In 1991, The Alliance of Small Island States (AOSIS)- Green Island Case studies (AOSIS, 1991) was formed as a coalition of small islands and low-lying coastal countries which face similar development challenges and considerations concerning the environment, particularly their vulnerability to the adverse effects of worldwide climate change. The region coverage includes: Atlantic, Baltic, Caribbean Indian Ocean, Mediterranean, Pacific, Southeast Asia (44 States). It capacities basically as an advertisement hoc campaign and negotiating voice for small island developing States (SIDS) among the international organization system United Nations. The sector covered were: general development problems, environment, climate change, global warming. Member States of AOSIS work along fundamentally through their New York diplomatic Missions to the United Nations. Major policy decisions are taken at ambassadorial-level plenary sessions.

What was accomplished: As a result of this initiative, studies were produced for 42 Small Island Development States to identify the most appropriate ones for applying sustainable energy practices

In the same year 1992, the United Nations Conference on Environment and Development (UNCED) took place with the aim to discuss the interlinked problems of environment and development and what the international community could do towards more sustainable pathways. The action plan of UNCED highlight the importance of developing strategies for policies and actions. Following this, in 1994, the UN Global Conference on Sustainable Development of Small Island Developing States which took place in Barbados, highlighted the principles and commitments to sustainable development made at UNCED and set out a programme of action. Following these, many countries have adopted more comprehensive national strategies and plans with the purpose of integrating environmental and development objectives (Bass & Dalal-Clayton, 1995). This conducted to a large number of experiences which lead to many lessons learnt. However, dependent territories tend to be heavily dependent on their mother countries. As a result, regional trade can be relatively slight.

In 2000, Global Sustainable Energy Islands Initiative (GSEII) has joined forces with all seven Caribbean island nations on the arrangement and execution of Sustainable Energy Plans (SEPs) and National Energy Policies (NEPs). Helping accomplice islands with NEPs and SEPs is GSEII's major strategy of expelling political obstructions to renewable energy and energy efficiency projects. Functioning on a national level, GSEII partners with islands governments to create and execute these policies and plans. GSEII has worked on numerous islands to write down arrangements and policies where none existed, upgrade and improve them. NEPs (the policies) are the goals and targets; for example, to increase the use of renewable energy to 20% of total energy production by 2020. The SEPs are the specific ways (the plan) that policies can be realized such as the government is to facilitate acquisition of land for the wind farm planned by the national utility on the east coast.

The plans have and proceed to recognize administrative and incentive measures, create targets and timetables for activity, energize the improvement of industry capability and accomplices for project plan and usage, and recognize the requirement for help in securing finance and investment sources.

What was accomplished: As a result of this initiative, it was launched the *Vision 2030: Partnership for Islands Economies* (United Nations) to scale up renewable energy in small island states. It also advocated and supported projects and partnerships for sustainable development while organizing and participating in several workshops, meetings and conferences on the topic.

Later in 2006, DAFNI network (Dafni.net), an innovative, nonprofit organization (NGO) and a covenant coordinator for Aegean islands which supports sustainable development in the Greek islands was formed. In this scheme, participate almost all the non-interconnected islands except Crete, including a number of those connected with the mainland (Aegina, Hydra, Kea, Andros, Samothrace), per total 3 regions/prefectures and 33 islands (Dafni.net). Every member of this community is committed to attain a number of objectives which are subject to external certification. These objectives can be summarized in the following sentences

- Economic development while taking into consideration environmental and social factors always with a first priority tourism
- Innovation in different forms which can be reflected principally in new renewable energy and water desalination projects.

Part of DAFNI scheme (Dafni.net), is being developed a smart grid system in collaboration with the ELENA programme, (which provides technical assistance), DG ENER, Covenant of Mayors and the European Investment Bank (EIB) which will provide financing for the project. This project includes the installation of smartelectricity meters, energy control centers, PVs and electric vehicle charging stations (EVCS) in the following NII: Lesvos,Limnos, Milos, Kithnos and Santorini

Also DAFNI network assisted in the creation of an e-learning platform with the objective to enhance energy planning and improve energy systems in the islands with the purpose of meeting the European environmental targets for 2020.

What was accomplished: As a result of this initiative, High Level Political meetings and declarations has been planned on topics related to sustainable development on islands. Also, acted in several projects as a mediator between the public (local authorities) and private sector in order to assist in sustainable development on Islands.

In 2008, the Covenant of Mayors for Climate and Energy (convenantofmayors.eu), a European movement interesting to increase the use of sustainable energy resources and energy efficiency cooperated directly with local and regional authorities in order to implement projects to achieve EU climate targets for 2020 (20-20-20) and 2030: 40% reduction in CO_2 emissions, along with the implementation of Sustainable Energy Action Plans (SEAPs). The following non interconnected island municipalities: Ios, Kea, Lipsi, Milos, Limnos, Nisyros, Syros and Skyros participate in the Covenant of Mayors.

The sectors covered were: Energy & Climate Action: decarbonization of the energy sector, adapt to climate change impact, adapt to unavoidable climate change impact

In 2009, The Pact of Islands is a European (dafni.net) initiative embraced by the European Parliament grasps European island authorities that commit to take action and undertake activities in line with the EU 2020 energy targets. These include increase of renewable energy, energy efficiency and sustainable transport projects at local level. The island authorities signing the Pact of Islands enter a political engagement that highlights the vulnerability of islands to climate change and energy security, and promotes the increase of renewable. This initiative suggests new energy and environmental policies with ultimate goal to find financing through revolving mechanisms in order to recycle financial resources and increase investments for sustainable systems. The pact of islands aims the use of revolving funds to access different subsidizing programmes such as: National Strategic Reference Framework (NSRF) and Joint European Support for Sustainable Investment in City Areas (JESSICA); The development of renewable energies in each island depending on the island's characteristics; The collaboration with international institutions, such as the International Renewable Energy Agency (IRENA) and the further extension of this programme in neighboring isle; The promotion of this action along with the other projects founded by EU through energy days, workshops, the media with the participation of the local communities. Member of the Pact of Islands (which is an instrument similar to the Covenant of Mayors) have become the following NII: Ios, Kythnos, Lesvos, Limnos, Lipsi, Milos, Mykonos, Naxos, Donousa, Ereikousa, Iraklia, Koufonisi, Santorini, Sifnos, Skyros and Syros and also the Regions of North and South Aegean.The Pact of islands is consisted of 64 island authorities and is operating in collaboration with the European Commission

What was accomplished: As a result of this initiative, 53 sustainable energy projects on islands promoted 12 of which were successfully completed. Also it manage in total to reduce emission by 25% from sustainable actions on 56 islands.

In 2012, Islands First organization (islandsfirst.org) has been a great support for the Chair of AOSIS, a consolidation of 44 low-lying island and coastal nations, which ambitious is to viably address climate change worldwide. It play an instrumental role in the creation of the United Nations Framework Convention on Climate Change (UNFCCC) and its Kyoto Protocol.

What was accomplished: As a result of this initiative, the following have been achieved: Secure the integration of the amendments to the Kyoto Protocol for the low lying 44 islands Launch of the Warsaw International Mechanism to assist susceptible countries to cope with inevitable impact of climate change

In the same year 2012, DIREKT – Small Developing Island Renewable Energy Knowledge and Technology Transfer Network (direct-project.eu) was established. This is a cooperation scheme including universities from Germany, Fiji, Mauritius, Barbados and Trinidad and Tobago, funded by ACP Science and Technology Programme, an EU programme for cooperation between the European Union and the ACP region (Africa, Caribbean, Pacific). Their main aim is to strength the science and innovation capacity in the field of renewable energy.

What was accomplished: As a result of this initiative, five pilot projects aiming at providing assistance for renewable energy and technology transfer initiated in the following islands: Barbados, Fiji (technology transfer pilot), Mauritius, Trinidad & Tobago

Another one is the ENNEREG - Regions paving the way for a Sustainable Energy Europe project (ENNEREG), which was supported by Europe under the Intelligent Energy - Europe programme and took place between 2010-2013. This project had the same objective to support European Regions to attain their EU 20-20-20 commitments. In order to become more efficient this wider project was split in 12 sub-projects for 12 Pioneer ENNEREG Regions which work on the production of regional Sustainable Energy Action Plans (SEAPs) and initiating Sustainable Energy Projects (SEPs). The ENNEREG worked in parallel with the above actions/programmes by the Regional Authority of South Aegean which also aimed to promote a more sustainable system for the Greek islands. The sectors covered are: Energy Efficiency in Buildings/ Industry Efficient Appliances Transporation Renewables Clever Education Energy Supervision and Monitoring

What was accomplished: As a result of this initiative, 300 Sustainable Energy Projects (SEPs) were supported with 250 million euros with many of them being on European isolated islands. More than 2 million tonnes of CO2 reduction.

2014 was a key year of bringing the global community together. World Bank launched the global program The Small Islands States Resilience Initiative (SISRI) (gfdrr.org), as a response to calls by Small Island States for greater and more effective support to build their resilience to climate change and disaster risk. SISRI aims to tackle the problem of the Disaster Risk Management and climate resilience, by providing technical assistance to overcome capacity challenges in terms of technical aspects of investments. Furthermore supports Small Island States in accessing scaled up and more efficient financing for resilience.

What was accomplished: As a result of this initiative, 93 sustainable projects (energy, water, land, management, resilience infrastructure, support against climate change impacts etc) in 23 islands states with total commitment of 888 million US$

This was followed by another initiative taken by IRENA who established a Global Renewable Energy Islands Network (GREIN) (2014-2016) to facilitate a 'pool' for concentrating knowledge, case studies, examples of other islands, innovations and other useful information for rapid development for renewable energy resources in different island areas of the world. The regions covered were: Islands in Pacific, Caribbean, West Africa Regions. The analysis focused on the following sectors: Power Grid Integration; Resource Assessment; Tourism Applications; Desalination; Waste to Energy

The results from the analysis of this knowledge will be used from all the islands worldwide but also from "virtual islands" (in India, Africa) distant from transmission grids which cannot be connected to the main transmission grid and used so far resources with high costs. The main targets if GREIN programme are to develop roadmaps for islands, at a second step to implement them and to strength GREIN structures.

What was accomplished: As a result of this initiative, one of the biggest achievement was to assist authorities in those islands to develop their national renewable energy roadmaps under the target to improve quality of life in these regions and assist in climate change mitigation. Some examples: Launch of the ECOWAS Centre for Renewable Energy and Energy Efficiency (ECREE) in Praia, Cape Verde; Energy framework and legislation in Cape Verde establishment; CARICOM Ministers of Energy published the Regional Energy Policy (REP) in 2013.

In 2015, Through the Energy Transition Initiative (ETI) (energy.gov), the U.S. Department of Energy and its partners US Islands work with government entities and other partners to set up a long-term energy vision and effectively actualize energy efficiency and renewable energy solutions. ETI gives a demonstrated system and specialized tools to offer assistance islands, states, and cities move to a clean energy economy and accomplish their clean energy objectives. It covered the following sectors: Renewable Energy; Energy Efficiency; Energy scenarios and projections; Data analysis; Extreme weather events (lessons learnt & management)

The initiative is divided in seven different actions:

1. Islands Playbook
2. Island Energy Snapshots
3. Island Energy Tools and Trainings
4. ETI Energy Scenario Tool
5. State and Local Energy Data Tool
6. Energy Resources for Hurricane Season
7. Energy Resources for Tornado Season

What was accomplished: As a result of this initiative, several things have happened:

- Hawaii published the target of Achieving 70% Clean Energy by 2030
- U.S. Virgin Islands approves the Interconnection Standards to Clear the Way for Grid Interconnection
- US Virgin Islands set the goal to reduce fossil fuel dependence by 60% by 2025
- Several initiatives at research and commercialization level supported across the islands with technical expertise and economic provision.

For example new technology to Find and Fill Building Energy Leaks, improved components for more efficient and clean diesel engines.

- Enchantment of solar water industry in Barbados
- Support of solar PV projects in Oahu and several other small and medium scale projects across the islands.

Small Island Developing States (SIDS) established in 2016 have played an essential part in accomplishing the Sustainable Development and Climate agendas for small island nations. The proposal consists in the creation of an Initiative for Renewable Island Energy (IRIE) with the purpose to support SIDS in their execution of the renewable energy and energy efficiency components of their nationally determined contributions (NDCs) to execute the Paris Agreement

What was accomplished: As a result of this initiative, $500 million were directed to new investments to implement 100 MW of new solar PV projects and 20 MW of wind between 2016-2021

More recently, The Smart Islands Initiative aims of putting the Smart Islands Initiative into action. On 28 March 2017 in Brussels, island local authorities sign up the Declaration, which aims to continue to be committed to a low carbon economy and inspire other communities in order to help environment. The sectors covered are: energy, electricity, transport, water and waste.

There are several conclusions to be drawn from the review of these initiatives. As seen in all of them, the development strategies were intensively focused. Some countries have a more specific development plan. This gives encouraging signals for private sector investment. However to attract private investors, global support is needed. The RE targets need to be harmonized between the PICTs own national RE targets and other targets (e.g. Sustainable Energy for All Initiative (SE4All). The SIDS Dock aims at generating a minimum of 50% of electric power from renewable energy sources by 2033. Another critical gap is the lack of measuring resiliency and management of resources under different environment extreme scenarios.

Initiatives such as Pact of Islands strengthen the solid public-private association required to advance RES and energy efficiency in the European Islands. Engaging the stakeholders, learn how to develop sustainable energy actions, twinning and replication are key lessons.

One of the key sector in most of the islands is tourism. Next section discusses practices towards sustainable tourism

PRACTICES TOWARDS SUSTAINABLE TOURISM

Under the context of a resilient, sustainable tourism industry a number of SIDS and other countries which occupy islands in their territories have attempted to establish policies and strategic plans to mitigate the impacts of tourism on the islands. It has to be underlined that the aim of the islanding relying on tourism is not to reduce the level of tourism with multiple socio economic benefits but rather to reduce the environmental impact (Briguglio L. & Brigulio M., 1996).

A number of proposed actions are summarized below:

1. **Long-Term Planning AND Regulation Framework:** The appropriate long term plans and strategies for islands with the community and stakeholder involvement, in parallel with concrete regulations are necessary to set the ground for monitoring tourism on islands (Sheldon, 2002; Trousdale, 1999). Although the plans have to take into account the local culture and the needs of the indigenous population, self-regulation was proved to be unsuccessful from the local stakeholder because of the profit opportunities e.g. structures on beaches. In that sense, common centralized (government intervention & legal controls) and decentralized efforts (community participation) are required to set the framework for long term planning on tourism industry development on islands (Briguglio, 2011; Sheldon, 2005). Malta exemplifies with the national Structure Plan which enabled mechanisms of monitoring and regulating tourism development on the island a long time ago.
2. **Preservation of the Island Culture:** It is very important for the islanders to preserve their local community values and culture as part of the wider development planning (Castri, 2002). Buildings and other structures of historical and cultural importance have to be preserved and not distorted from the touristic infrastructure. Furthermore, local traditions must be well-kept-up and empowered . Although this point, is not directly related with the environmental impact of tourism, respect to the local heritage and community is of paramount importance in order to set a solid framework for sustainability in tourism
3. **Environmental Management:** Due to the fact that the islands have small, fragile eco-systems usually with rich bio-diversity Environmental Impact Assessments Planning is crucial for establishing rules, standards and monitoring environmental management on the islands (Briguglio

& Brigulio, 1996). Firstly, land use management has to be regulated with robust policies which will allow a rational allocation of the land uses among tourism, agriculture and other industries. Continuing, the preservation of the natural environment could become part of the visiting experience, for example in Seychelles a number of activities associated with the local wildlife are offered to tourists (Shah, 2002).

Apart from the land management, strict regulations for the coastal areas and the ocean are necessary. However, it is important the combination of strong policies and the active participation of the local community and the tourists in this effort through actions such as recycling and waste management treatment. As in some islands there are not sufficient landfills, the case of transporting waste to the mainland used to be an option for some islands (Trousdale, 1999), however with considerable economic and environmental implications. As such, smart ways of dealing with wastes must be adopted on the islands such as waste treatment to biofuels etc for energy production.

In addition, energy conservation in commercial buildings and tourism facilities as well as use of alternative energy fuels either at a large or small scale are essential towards sustainability in tourism. Special incentives and regulations for the islanders and the local stakeholders must be provided by the local or the central government to accelerate renewable energy development. Also, in terms of energy efficiency it is important to set standards and monitor the performance of both medium and large accommodation facilities or commercial, recreational centres (Manning, 2016).

At this point is important to differentiate the practises of large scale sustainable resorts with high sustainable management levels and smaller chains or local business which usually are in a disadvantage and don't or cannot apply these practises unless there is a strong incentive in term of sacrificing their provided services offered to their customers. Also, it is claimed that instruments such as eco-labelling might not be efficient enough to produce actual results (de-Miguel-Molina et al., 2014).

4. **Monetization:** On several vulnerable islands environmental management comprises recovery from natural disasters with severe implications on the local built and natural environment (Meheux & Parker, 2006). The required funds for conservation and disaster management could be raised through special taxes, subsidies additional fees to the tourists or other instruments (Shah, 2002). Monetizing the environmental impact could raise the awareness that both climate change and environmental

pollution may have a considerable socioeconomic cost, even though this is not usually reflected in the market prices (Briguglio & Brigulio, 1996).

5. **Visitor Management:** Visitor management could be integrated in the section of long term planning although it requires additional consideration. Monitoring the number of tourists arriving as well as other characteristics of their visits such as duration of their stay, profiles of the visitors (age, gender, preferences, transport means etc) is important in terms of a careful planning. On the basis of a long-range planning is very important to monitor and record the balance of supply and demand of tourism, at a quantitative and qualitative level in order to prepare for the following years and take the appropriate measures for lowering tourism impact on the various sectors. A study of tourism for the Canary Islands proved that when supply and demand are not well balanced the tourism industry will encounter challenges and complications (Gil, 2003).
6. **Information Systems and Knowledge Management:** The necessary knowledge and information systems could facilitate as tools for assisting islands to promote their tourism product and in parallel manage their tourist resources. Online tools for advertising the islands, as well web-based tools for statistics analysis and marketing could assist islands on building a robust contemporary market industry based on tourism. Furthermore, smartphone applications and clever on-line tools based on data frameworks, geographical data frameworks mapping frameworks (GIS) as well worldwide situating frameworks (GPS), and brilliantly transportation frameworks could further improved touristic services while in parallel allowing for a better monitoring of the tourists activities on the islands. Similar GIS systems, are used for spatio analysis to monitor the environmental impact on land, air and sea, an example has been demonstrated on the Cayman islands with a tool to evaluate tourism impact on fragile reefs (Hall, 1998) .
7. **Accessibility and Transportation:** Transportation to and on the islands is an area where drastic changes could take place mainly in terms of the transport sector on the islands. As usually, tourists use their own vehicles due to lack of sufficient public transport means on the islands, there is no adequate monitoring of their performance, consumption and emissions (Sheldon, 2005). However, new practises in terms of car-sharing, electro mobility and hydrogen, fuel cell cars could be applied successfully on small sized islands at a demonstration level first and secondly at a commercial level.

Special consideration deserves the ocean and air pollution from transportation to the islands. In that case, there is limited action and mainly stemming at a central governmental level based on the level of usage of the various transport means, e.g. ships, cruise ships, airplanes etc. For example, in the Hawaiian Islands a high-speed ferry has been placed among the islands to offer supplementary capacity to the inter-island flights (Sheldon, 2005).

8. **Alternative Tourism and Market Divergence:** The quality of tourists visiting an island and the quality of the touristic product are significant for islands. Niche markets on the islands could result in the highest expenditure for the island, as took place in the Canary Islands (Diaz-Perez et al., 2005). Hawaiian Islands attempted to pinpoint niche opportunities and generated the Hawaii's Strategic Plan 2005-2010 setting a methodology on assessing the quality of tourists while placing emphasis on the expenditures and other characteristics associated with their activities on the islands.

Although most of the islands become popular for their natural beauty, the good weather and their luxurious facilities, it is very important for a number of islands to develop niche markets and attract alterative forms of tourism such as eco-tourism and cultural tourism with usually reduced environmental footprint for the island (Briguglio, 2011). Another, indirect benefit is that this kind of tourism distributes visitors among different places of the island avoiding their massive concentration in the coastal areas.

The development of speciality markets based on the islands resources is an important market strategy. For instance in Tasmania, Australia, the local stakeholders decided to promote agricultural based tourist activities for example accommodation in the farms or agricultural museums, wine tasking, as well as other activities such as lavender and raspberries crops collection (Conlin, 2002).

CASE STUDIES

Caribbean Islands

The Caribbean islands constitute one of the most popular and exquisite vacation destinations in the world with a unique landscape where the projection show that tourism will continue its growth over the next years. The Caribbean

Tourism Organization has put into the forefront a number of sector that need to be supported such as maintenance of product quality without compromising the natural environment but rather integrating it. Also access to the islands by air with competitive prices, support of initiatives and interlinkages among tourism and other sectors in terms of economic development and marketing (Goodwin, 2008; UNEP Islands, 1996).

To become more explicit the Caribbean islands have already acted towards this direction with the establishment of the following instruments:

The Sustainable Destinations Alliance for the Americas (SDAA) established in 2014 which is the first-ever large-scale multi-sector programme focused to sustainable tourism goals in the Caribbean and Latin American regions. This initiative has an objective to support tourism in the region through sustainable forms while in parallel empowering the competitiveness as touristic destinations. The initiative assists in the management and marketing of the islands while providing different tools for measuring and analyzing goal sustainability will be given, multi-stakeholder goal management will be empowered and the local capacity inclusion.

In 2013 the Greater Caribbean region was presented as the world's first "Sustainable Tourism Zone", in order to protect and ensure tourism as a long-term, sustainable and economic profitable activity is one of the principal concerns while trying to establish the Caribbean region as a Sustainable Tourism Zone.

Finally, the Caribbean Hotel Energy Efficiency and Renewable Energy Action Programme in organization with the Caribbean Hotel and Tourism Association is the last and equally important initiative with pilot projects applied in Barbados, Bahamas and Jamaica. This programme will contribute in the configuration of future policies related to environment and energy. This initiative aims to support energy conservation and the use of alternatives forms of energy in the Caribbean hotel sector. Morevover, it searches for incentives and support schemes to motivate hoteliers in the region to apply practices for reducing the carbon footprint of the sector.

Malta and Cyprus

Malta and Cyprus belonging to the European Union are committed in a number of targets for decarbonizing their industry which relies on a high level on tourism. It is noticeable that in those two Island States there was recorded a slower pace of growth in tourism in the previous decade compared to the older trends, as a consequence of the capacity constraints which was an outcome of a number of policies towards environmental conservation. Both islands states, although they rely significantly on tourism and they aim to continue promoting and marketing the countries as touristic destinations mainly as up-market tourism, they established a number of regulation towards a controllable expansion of the touristic facilities on the islands while they focused on the coastal areas (UNEP Islands, 1996).

Galapagos Islands

Galapagos islands through the Galapagos National Park Directorate (GNPD) and the Ecuadorian Ministry of Tourism have targeted in the previous years on transforming the local tourism sector into an environmentally friendly sector. The main pillars for action on the islands are: water, energy and waste while they focus on enhancing the local production. Also training of the local population on concepts of sustainable tourism, marketing etc. The local authorities in collaboration with other private stakeholders encourage tourists to pay attention on Galapagos conservation while using experienced, professional travel agencies and providers with social responsibility (Sustainable Tourism in Galapagos, n.d.).

Easter Island

Tourists visiting the Easter Island are encouraged of the unique and fragile natural environment of the island including was to avoid its disturbance through informative sessions, leaflets and other ways in order to show to the tourists the concept of sustainability on the island. They try to focus on particular areas and sight sighting which concentrate attention while in parallel empowering local production of goods and products (Baillargeon, 2015).

El Hierro

The island of El Hierro is well known for its practises on the energy sector with the development of the hydro pump wind power station achieving RES penetration to 100% in the electricity sector. In parallel, it was declared a Biosphere Reserve by the UNESCO and supported sustainable development practices in the tourism sector promoting the concept of Sustainable and Alternative Tourism while demonstrating the cultural and environmental heritage of the island through campaigns (The island of El Hierro, a model of sustainable Tourism, n.d.).

Roatan (Honduras)

Roatan has been on the spot of an island with increasing popularity while becoming a tourism hotspot. Only in 2015 more than 1 million travellers visited the island. The island including the local tourism industry and community is very active on supporting sustainability as the last 4 years the Roatan Geotourism Stewardship Council, the Go Blue Central America Geotourism MapGuide, the Coral Reef Alliance and the Mesoamerican Reef Tourism Initiative (MARTI) have provided training courses to hundreds of local and foreign stakeholders and business in terms of in sustainable tourism best practices (Rainforest Alliance, 2016).

FUTURE RESEARCH DIRECTIONS

Around the world, national energy policies have been developed and practical strategic action plans to implement the policies have been adopted. In addition, commonsense instruments to execute the activity plans have been created. Strategic planning initiatives and reforms will definitely help to identify the ways in which policy and regulation block the vision of islands and find solutions. Through projects in the ground, education, increased connectivity and quality of life, we can develop a detailed understanding of a future energy system in islands. Regulation and policy is a key barrier to accessing finance and funding, and regulatory change needed to enable more strategic investment in infrastructure.

There is an increased visibility and understanding of key energy issues. There is an increased build up participation coordination and trade of information across networks. Such effective illustrations are: Cook Islands Energy Action Plan (CIEAP), which was created in 2005 with accentuation on energy security, energy conservation and renewable energy development and The Action Plan (NEAP) which have been endorsed by Cabinets in 2006. All these efforts are a step forward.

For tourism, transport sector could play a key role in reducing carbon emissions. Possible solutions may consist in promotion of the use of bicycles and electric bikes to navigate the island, use of electric scooters, replacement of diesel cars for rent by electric or plug-in hybrids, introduction of electric, hybrid or biofuel-powered buses. However, most of these solutions require the development of a network for low-power charging points. In the case of islands with marina, water transport could be equipped with hybrid propulsion. In the case of islands which are not interconnected with the national power system, and operate in an island mode, charging points should be equipped with demand response communication and control systems to reduce power consumption. Vehicle-to-Grid (V2G) when energy flows both to and from the vehicle, turning in into a portable battery store could be a feasible option for islands. This will help balance the electricity grid by providing energy from their vehicle batteries to the grid at times of peak demand and benefit from rewards based on the provision of grid services and energy trading revenue.

CONCLUSION

This chapter exploits the information on peculiar challenges faced by islands nations with respect to sustainability, the issues raised by these challenges, the reaction in terms of projects and policy initiatives to cope with these challenges and identification of sectors for paving a sustainable future of island nations.

The analysis illustrates and confirms that among island societies, the sensation of being before a new scenario went beyond the technical level, and turned to be one of the main debates on islands' options of future. Several declarations such as the one of the Canary Islands (1998), Palma de Majorca (1999), Azores (2000), Cagliari (2001) and, especially, the one of Chania (Crete -2001), as well as meetings of large transcendence such as the one held by the Small Island Developing States on Mauritious (BPoA+10 - 2005), which stressed on the crucial role of renewables in the consolidation of actual sustainable development strategies and their relation with the improvement of economic development and quality of life of island populations. Islands have, therefore, reached an important moment in their evolution.

The decarbonisation plans will be achieved through a basket of solutions, such as renewable energy integration, energy efficiency, transport efficiency, use of information communication technology. Some islands have set the scene for offshore wind farms, others for solar, biogas and so on. However, to determine which renewable energy systems are most suitable, a great understanding of current situation in islands is needed. Therefore the information highlighted in this chapter will give an indication of political will and possible technical solutions for key sectors, making it advisable to open the gates wide to this great idea of energy self-sufficiency for islands in the new millennium.

REFERENCES

Ayto, J. (1990). *Arcade Dictionary of Word Origins the Histories of more Than 8,000 English-Language Words*. Arcade Publishing.

Baillargeon, Z. (2015). *Sustainable Tourism on Eastern Island*. Retrieved from https://www.cascada.travel/en/News/Sustainable-Tourism-Easter-Island

Baric, D., Bredin, D., Dressler, C., Jansson, M., Kechagioglou, E., & Lodwick, N. (2016). *Atlas of the small European Islands*. ESIN European Small Islands Federation. Retrieved from https://europeansmallislands.files.wordpress.com/2014/11/atlas-of-the-esin-islands-27-91.pdf

Bass, S., & Dalal-Clayton, B. (1995). *Small island states and sustainable development: strategic issues and experience*. Environmental Planning Issues, No. 8. International Institute for Environment and Development, UK. Retrieved from http://pubs.iied.org/pdfs/7755IIED.pdf

Beate, M. W. R. (2018). Geography of Small Islands. Springer International Publishing.

Blanco, G., Gerlagh, R., Suh, S., Barrett, J., de Coninck, H. C., Diaz Moreion, C. F., ... Zhou, P. (2014). Drivers, Trends and Mitigation. In O. Edenhofer, R. Pichs-Madruga, Y. Sokona, E. Farahani, S. Kadner, K. Seyboth, ... J. C. Minx (Eds.), *Climate Change 2014: Mitigation of Climate Change. Contribution of Working Group III to the Fifth Assessment Report of the Intergovernmental Panel on Climate Change*. Cambridge, UK: Cambridge University Press.

Briguglio, L. (2011). *Sustainble tourism on small islands with special reference to Malta*. Malta.

Briguglio, L., & Brigulio, M. (1996). Sustainable tourism in small islands: The Case of Malta. *Sustainable Tourism in Islands and Small States: Case Studies*, (1993), 162–179.

Burnham, A. D., & Berry, A. J. (2017). Formation of Hadean granites by melting of igneous crust. *Nature Geoscience*, *10*, 457–461. doi:10.1038/ngeo2942

Cayman Islands Government. (2013). *National Energy Policy 2017-2037 Cayman Islands.* Retrieved from http://www.gov.ky/portal/pls/portal/docs/1/12374582.PDF

Church, J. A., Clark, P. U., Cazenave, A., Gregory, J. M., Jevrejeva, S., Levermann, A., ... Unnikrishnan, A. S. (2013). Sea Level Change. In *Climate Change 2013: The Physical Science Basis. Contribution of Working Group I to the Fifth Assessment Report of the Intergovernmental Panel on Climate Change.* Cambridge, UK: Cambridge University Press.

Conlin, M. (2002). *Tasmania: Balancing Commercial and Ecological interests in tourism development.* Leiden: Biodiversity and Information.

Conrad, M. D., Esterly, S., Bodell, T., & Jones, T. (2013). *American Samoa: Energy Strategies.* The National Renewable Energy Laboratory (NREL) Report for U.S. Department of the Interior's Office of Insular Affairs (OIA). Retrieved from http://www.covenantofmayors.eu/en/

DAFNI Network of Sustainable Greek Islands. (n.d.b). *ELENA Technical Assistance awarded to DAFNI network – for preparing a project for Smart Grids in 5 Aegean Islands.* Retrieved from http://www.dafni.net.gr/en/collaborations/elena-ta.htm

DAFNI Network of Sustainable Greek Islands. (n.d.c). *Sustainable Energy and European Islands: The Pact of Islands and the ISLE-PACT project.* Retrieved from http://www.dafni.net.gr/en/collaborations/pact-of-islands.htm

DAFNI Network of Sustainable Greek Islands. (n.d.a). Retrieved from http://www.dafni.net.gr/en/home.htm

Darwin, C. (1859). *On the Origin of Species.* New York: D. Appleton and Company. Retrieved from: http://darwin-online.org.uk/converted/pdf/1861_OriginNY_F382.pdf

de-Miguel-Molina, B., de-Miguel-Molina, M., & Rumiche-Sosa, M. E. (2014). Luxury sustainable tourism in Small Island Developing States surrounded by coral reefs. *Ocean and Coastal Management, 98*, 86–94. doi:10.1016/j.ocecoaman.2014.06.017

Depraetere, C. (2008). The Challenge of Nissology: A Global Outlook on the World Archipelago Part I: Scene Setting the World Archipelago. *Island Studies Journal, 3*(1), 3–16.

Depraetere, C., & Dahl, A. L. (2007). Island Locations and Classifications. In G. Baldacchino (Ed.), *A World of Islands* (pp. 57–105). Charlottetown: Institute of Island Studies, University of Prince Edward Island in collaboration with Agenda Academic, Malta.

Di Castri, F. (2002). *Diversification, Connectivity and Local Empowerment for Tourism Sustainability in South Pacific Islands – a Network from French Polynesia to Easter Island.* Leiden: Biodiversity and Information.

Diaz-Perez, F., Bethencourt-Cejas, M., & Alvarez-Gonzales, J. (2005). The segmentation of Canary Island tourism markets by expenditure: Implications for tourism policy. *Tourism Management*, *26*(6), 961–964.

DIREKT – Small Developing Island Renewable Energy Knowledge and Technology Transfer Network. (n.d.). Retrieved from http://www.direkt-project.eu/objectives.html

ENNEREG – Regions paving the way for a Sustainable Energy Europe Project. (n.d.). Retrieved from https://ec.europa.eu/energy/intelligent/projects/en/projects/ennereg

Eurelectric. (2012). *EU Islands : Towards a Sustainable Energy Future.* Retrieved from https://www3.eurelectric.org/media/38999/eu_islands_-_towards_a_sustainable_energy_future_-_eurelectric_report_final-2012-190-0001-01-e.pdf

Forbes, D. L., James, T. S., Sutherland, M., & Nichols, S. E. (2013). Physical basis of coastal adaptation on tropical small islands. *Sustainability Science*, *8*(3), 327–344.

Gil, S. (2003). Tourism Development in the Canary Islands. *Annals of Tourism Research*, *30*(3), 744–747.

GIZ. (2016). *Coping with Climate Change in the Pacific Island Region.* Retrieved from https://www.giz.de/en/worldwide/14200.html

Global Sustainable Energy Islands Initiative (GSEII). (n.d.). Retrieved from http://gseii.org/

Goodwin, J. (2008). Sustainable Tourism Development in the Carribean Island Nation-States. *Michigan Journal of Public Affairs*, *5*, 1–16.

Government of the Republic of Kiribati. (2009). *Kiribati National Energy Policy.* Retrieved from http://www.mfed.gov.ki/sites/default/files/Kiribati%20 National%20Energy%20Policy.pdf

Government of the Virgin Islands. (2016). *Energy Policy, Ministry of Communications and Works.* Retrieved from http://www.bvi.gov.vg/sites/default/files/resources/energy_policy_of_the_virgin_islands_oct_2016.pdf

Grigor'yev, G.N. (2014). A Genetic Classification of Islands. *Soviet Geography, 12*(9), 585-592. DOI 10.1080/00385417.1971.10770277

Hawaii State Energy Office. (n.d.). *State of Hawaii- Energy policy directives.* Retrieved from http://energy.hawaii.gov/energypolicy

IRENA. (n.d.). *A Global Renewable Energy Islands Network (GREIN).* Retrieved from https://sustainabledevelopment.un.org/partnership/?p=8011

Islands First. (n.d.). *On the whole the impacts of climate change on small islands will have serious negative effects…* IPCC Fifth Assessment Report. Retrieved from http://www.islandsfirst.org/climate-change

Kaldellis, J. K., Gkikaki, A., Kaldelli, E., & Kapsali, M. (2012). Investigating the energy autonomy of very small non-interconnected islands. A case study: Agathonisi, Greece. *Energy for Sustainable Development, 16*(4), 476–485.

Kasselouri, B., Kambezidis, H., Konidari, P., & Zevgolis, D. (2011). Environmental, economic and social aspects of the electrification on the non-interconnected islands of the Aegean Sea. *Energy Procedia, 6*, 477–486.

Maldives Energy Authority. (n.d.). *Legal and regulatory framework of the energy sector of the Maldives.* Retrieved from http://www.irena.org/EventDocs/Maldives/2LegalRegFrameworkMaldivesEnSectorEnAuth.pdf

Manning, E. W. (2016). *The Challenge of Sustainable Tourism in Small Island Developing States (SIDS).* Retrieved from http://canadiancor.com/challenge-sustainable-tourism-small-island-developing-states-sids/

Marin, C., Alves, L. M., & Zervos, A. (Eds.). (2005). 100% RES. A challenge for Island Sustainable Development. Instituto Superior Tecnico Research Group on Energy and Sustainable Development.

Meheux, K., & Parker, E. (2006). Tourist sector perceptions of natural hazards in Vanuatu and the implications for a small island developing state. *Tourism Management, 27*, 69–85. doi:10.1016/j.toutman.2004.07.009

Ministry of Housing and Environment. (2010). *Maldives National energy policy & strategy*. Retrieved from www.mhte.gov.mv

Ministry of Mines, Energy, and Rural Electrification. (2014). *Solomon Islands National Energy Policy*. Retrieved from http://prdrse4all.spc.int/system/files/volume1_solomon_islands_national_energy_policy.pdf

NASA. (2016). *Sea Level*. Retrieved from https://climate.nasa.gov/vital-signs/sea-level/ and https://www.nasa.gov/press-release/nasa-noaa-data-show-2016-warmest-year-on-record-globally

Niedzielski, T. (2014). Chapter Two – El Niño/Southern Oscillation and Selected Environmental Consequences. *Advances in Geophysics*, *55*, 77–122.

NOAA. (2016). *Is Sea Level Rising?* Retrieved from http://oceanservice.noaa.gov/facts/sealevel.html

NREL. (2015a). *Energy Transition Initiative-Energy Snapshot Barbados*. Retrieved from https://www.nrel.gov/docs/fy15osti/64118.pdf

NREL. (2015b). *Energy Transition Initiative-Energy Snapshot U.S. Virgin Islands*. Retrieved from https://www.nrel.gov/docs/fy15osti/62701.pdf

NREL. (2015c). *Energy Transition Initiative-Energy Snapshot Puerto Rico*. Retrieved from https://www.nrel.gov/docs/fy15osti/62708.pdf

NREL. (2015d). *Energy Transition Initiative-Energy Snapshot Jamaica*. Retrieved from: https://www.nrel.gov/docs/fy15osti/63945.pdf

NREL. (2015e). *Energy Transition Initiative-Energy Snapshot Micronesia*. Retrieved from: https://www.nrel.gov/docs/fy15osti/64294.pdf

NREL. (2015f). *Energy Transition Initiative-Energy Snapshot Curacao*. Retrieved from: https://www.nrel.gov/docs/fy15osti/64120.pdf

NREL. (2015g). *Energy Transition Initiative-Energy Snapshot -Dominican Republic*. NREL.

NREL. (2015h). *Energy Transition Initiative-Energy Snapshot Palau*. Retrieved from https://www.nrel.gov/docs/fy15osti/64291.pdf

Nurse, L. A., McLean, R. F., Agard, J., Briguglio, L. P., Duvat-Magnan, V., Pelesikoti, N., ... Webb, A. (2014). Small islands. In *Climate Change 2014: Impacts, Adaptation, and Vulnerability. Part B: Regional Aspects. Contribution of Working Group II to the Fifth Assessment Report of the Intergovernmental Panel on Climate Change* (pp. 1613–1654). Cambridge, UK: Cambridge University Press.

O'Brien, H. K. (2004). *Tokelau National Energy Policy and strategic action planning (NEPSAP).* Retrieved from https://www.tokelau.org.nz/site/tokelau/files/NEPSAP%20Main%20Text.pdf

Office of Energy Efficiency & Renewable Energy. (n.d.). *Energy Transition Initiative (ETI).* Retrieved from https://energy.gov/eere/about-us/energy-transition-initiative

Pacific Power Association. (2006). *United States of America Insular Areas Assessment Report: An Update of the 1982 Territorial Energy Assessment.* Prepared for US. Department of Interior Washington, Prepared by Pacific Power Association Suva, Fiji Retrieved from http://prdrse4all.spc.int/system/files/us_energy_assessment_-_rmi_fsm_palau_2006.pdf

Rainforest Alliance. (2016). *Sustainable Tourism Efforts Gain Momentum On The Island Of Roatan, Honduras.* Retrieved from: https://www.rainforest-alliance.org/articles/sustainable-tourism-efforts-gain-momentum-on-the-island-of-roatan-honduras

Sanseverino, E. R., Sanseverino, R. R., Favuzza, S., & Vaccaro, V. (2014). Near Zero Energy Islands in the Meditteranean: Supporting Policies and Local Obstacles. *Energy Policy*, *66*, 592–602.

Shah, N. J. (2002). Bikini and Biodiversity: Tourism and Conservation on Cousin Island, Seychelles. In F. di Castri & V. Balaji (Eds.), *Tourism, Biodiversity and Information* (pp. 185–196). Leiden, The Netherlands: Backhuys Publishers.

Sheldon, P. (2002). Information Technology Contributions to Biodiversity in Tourism: the Case of Hawaii, in eds. di Castri and Balaji. In Tourism, Biodiversity and Information. Backhyus Publishers.

Sheldon, P. J. (2005). The Challenges to Sustainability in Island Tourism. *Occasional Paper, 2005*(October), 1. Retrieved from: http://citeseerx.ist.psu.edu/viewdoc/download?doi=10.1.1.502.5607&rep=rep1&type=pdf

Small Island Developing States (SIDS). (n.d.). Retrieved from https://irenanewsroom.org/2016/11/17/scaling-up-renewable-energy-on-small-island-developing-states-the-initiative-for-renewable-island-energy/

Spilanis, I., Kizos, T., Biggi, M., Vaitis, M., Kokkoris, G., Lekakou, M., . . . Azzopardi, R. M. (2013). *The Development of the Islands – European Islands and Cohesion Policy (EUROISLANDS) Targeted Analysis 2013/2/2 - Interim Report* (version 3). Retrieved from: https://www.espon.eu/sites/default/files/attachments/INTERIM_REPORT_50510.pdf

The Alliance of Small Island States (AOSIS). (1991). Retrieved from http://www.globalislands.net/greenislands/http://aosis.org/

The Smart Islands Initiative Brussels. (n.d.). Retrieved from http://www.scottish-islands-federation.co.uk/the-smart-island-initiative/

Timilsina, G. R., & Shah, K. U. (2016). Filling the gaps: Policy supports and interventions for scaling up renewable energy development in Small Island Developing States. *Energy Policy*, *98*, 653–662. doi:10.1016/j.enpol.2016.02.028i

Trousdale, W. J. (1999). Governance in Context: Boracay Island, Philippines. *Annals of Tourism Research*, *26*(4), 840–867.

TUI Group. (2016). *TUI Holiday Atlas 2016*. Retrieved from https://www.tuigroup.com/en-en/media/stories/special-themed-section/christmas-new-year/2016-12-09-holiday-atlas

UN Country Team in Maldives. (2005). *Maldives Situation Report#20/2004*. Retrieved from https://reliefweb.int/report/maldives/maldives-situation-report-202004

UNEP & UNWTO. (2005). *Making Tourism More Sustainable - A Guide for Policy Makers*. Authors.

UNEP Islands. (1996). *Sustainable tourism development in small island developing States, Document E/CN.17/1996/20/Add.3, Commission on sustainable development Fourth session, 18 April - 3 May 1996*. Retrieved from http://islands.unep.ch/d96-20a3.htm

United Nations. (n.d.). *Vision20/30 – Partnership for Island Economies*. Retrieved from https://sustainabledevelopment.un.org/partnership/?p=759

UNWTO. (2013). *Réunion Island Declaration on Sustainable Tourism in Islands.* Retrieved from: https://sustainabledevelopment.un.org/content/documents/5219269reunion_declaration_final_en_0.pdf

Wallace, A. R. (2012). *Island Life: Or, The Phenomena and Causes of Insular Faunas and Floras, Including a Revision and Attempted Solution of the Problem of Geological Climates.* Cambridge University Press.

World Bank. (2016). *Climate Change in the Maldives.* Retrieved from http://web.worldbank.org/WBSITE/EXTERNAL/COUNTRIES/SOUTHASIAEXT/0,contentMDK:22413695~pagePK:146736~piPK:146830~theSitePK:223547,00.htm

World Bank Group GFDRR. (n.d.). *The Small Islands States Resilience Initiative (SISRI).* Retrieved from https://www.gfdrr.org/sites/default/files/publication/SISRI.pdf

Zafeiratou, E., & Spataru, C. (2017). Potential economic and environmental benefits from the interconnection of the Greek islands. *International Journal of Global Warming, 13*(3-4), 426–458. doi:10.1504/IJGW.2017.087211

Chapter 2
Interconnections and Islands:
Global Experiences and Learning

ABSTRACT

The goal of this chapter is to present the current state and technologies with regards to interconnections of islands to the mainland or between islands. The majority of islands interconnections have been recorded in Asia, although the longest projects are found in Europe. In Asia, the islands are usually located close to the shore 10-55 km or in island complexes such as Indonesia and Philippines where the enhancement of the national grids through interconnections with then neighboring islands is relatively easy achieved through short HVAC links. In Europe, longer projects are observed exceeding the 400 km mainly in the Mediterranean basin, while the new HVDC interconnections are expected to reach even longer lengths. In North America, only a few island interconnections have been implemented in close distances from the shore. The main driver to interconnect islands has been principally the requirement to access cheap energy sources usually located in the mainland.

INTRODUCTION

Power networks interconnections have played a key part in the history of electric power systems. We are referring in here at electric interconnections between islands or between mainland and one (or more) isolated islands. The electricity grids of isolated islands have a more fragile structure than

DOI: 10.4018/978-1-5225-6002-9.ch002

mainland because of lower number of generation sites, absence or insufficient interconnection and so on. Integrating large proportions of renewable energies that are volatile (such as wind and solar energy conversion systems) will require a well design interconnected system for insular nations. This chapter presents an overview of historical and current developments in terms of cable technologies, their reliability, risks and costs, of the insular power system structures including high RES penetration, of the interconnections plans, their challenges and opportunities, emphasizing on real examples from different insular areas across the globe.

There are few studies who focus on different topics related to islands. Diagne et al. (2013) provides a review of solar forecasting methods for insular power system operations. Brito et al. (2014) discusses the pros and cons of integrating sustainable energy resources for an imaginary insular power system. Interconnection of islands with the mainland together with increase of renewable energy systems in specific Greek cases have been examined from by Georgiou et al. (2011) from the point of technical applicability. Notton (2015) analysed the impact of grid integration of renewable energy in islands power systems and then analysed the current status of eleven French islands. These studies provided great insights into the challenges faced by islands. However, they do not take the problem in a more general framework considering all three factors economy, sustainability and reliability. Reviewing and understanding the evolution of technologies underpinned the integration of renewable energy and interconnection is key to understanding the future.

CABLE TECHNOLOGIES

Submarine Cable Technologies

The two technologies used in power transmission links are the HVAC (High Voltage Alternating Current) and the HVDC (High Voltage, Direct Current) usually used for longer distances and for systems interconnection. High Voltage (HV, including here also Extra-High Voltage – EHV – and Ultra-High Voltage – UHV) spans through the range of 35 kV – 800 kV which is expect to be expanded in the future (EuroAsia interconnector, 2013; Ardelean & Minnebo, 2015).

1. **High Voltage, Alternative Current (HVAC):** HVAC is utilized for wind offshore projects and islands interconnection, basically in near distances from the shore, as AC cables restrain the transmission distance for submarine cables to the break-even-length of 80 km (Ruddy et al., 2015). This is due to the capacitance between the active conductors and the adjoining sea-bed or water which constrains the length of the HVAC cables and limits its transmission efficiency as after certain lengths the reactive power used by the cable would take in the whole current carrying capacity of the conductor and no utilizable power would be transferred (Subsea Cables UK, 2013; Ardelean & Minnebo, 2015), . Be that as it may, this could be generally eased by positioning shunt reactors at the end of the cable (Gaji et al., 2003).

HVAC is converted to DC for transmission through the cable and it reverts to AC at the other end of the two interconnection points. AC connections appear higher levels of constancy compared to HVDC technology. They show 30% less recurrence in the events of 'expectable inability to supplied power (Daniel et al., 2014). The AC interconnection of an offshore wind farm comprises of the submarine transmission cables, two transformers offshore and onshore, reactive power compensators and the offshore platform. These days, submarine cables utilize for the most part expelled polymer (XLPE) insulation cover with copper or aluminum conductors, while three-core AC cables are chosen for submarine applications (NordPoolSpot, 2017).

2. **High Voltage Direct Current (HVDC):** The ongoing increase of offshore renewable energy as well as the requirement to interconnect energy isolated areas drives the underpinning of submarine infrastructure due to the requirement to transmit large power flows over long distances in a reliable and efficient way. As such, high voltage direct current (HVDC), is tending to ended up around the world the most reliable technology in submarine interconnections. In differentiate with AC, DC technology is not troubled by cable charging currents and permits mass control transmission through long distances while the transmissions lengths are almost limitless because of the abolition of the capacitive currents (Reidy and Watson, 2005). In addition, while an HVAC cable system requires three cables, a HVDC cable system requires no more than two.

Primarily the voltage source converter (VSC) HVDC technology permits fast control over dynamic and receptive control on the entirety operation scale through the AC-DC-AC converters able to meet and surpass all interconnection voltage/frequency control prerequisites. Additionally, HVDC can be associated to weaker control frameworks compared to AC interconnections permitting larger wind farm integration (Bresesti et al., 2007; Joveic & Strachan, 2009).

The primary shortcoming of DC cables is related to limited redundancy. That implies that an blackout in one pole leads to the add up to loss of the VSC-HVDC link. For that reason, the bipolar configuration is more apt for island and wind offshore interconnectors, since as only 50% of the transmission capacity could be lost in case of damage and set off to N-1 conditions (Nanou et al., 2014; Zafeiratou & Spataru, 2016).

Currently there are approximately 8000 km of HVDC submarine power cables world-wide. Nonetheless, this is a mere size compared to the total length of cables that are placed at the seabed which is equal to the astounding number of 10 million km mainly telecommunication cables. However, the rates for new HVDC submarine transmission installations are very promising and they will become rapidly an ever-present part of the electricity network transmission landscape (Subsea Cables UK, 2013).

3. **HVDC Light:** HVDC Light was initially introduced in 1997 with the first commercial HVDC Light cable system being installed in 1999 to link a wind park at the Gotland Island, and it is based on Voltage Source Converter transmission technology (Papadopoulos et al., 2008). HVDC light cables have been operational for almost 20 years, enabling power transmission of over long distances with lower costs and increased efficiency for submarine interconnections. The new HVDC Light cables are usually insulated with extruded polymer technology. The robustness and flexibility demonstrate the HVDC Light cables as appropriate cables for severe installation environment in submarine interconnections (Asplund et al., 2000). At this moment, HVDC light technology could be applied with technical features up to 300 kV and 1800 MW (ABB, 2006).

HVDC Light converters contain Insulated Gate Bipolar Transistors while they work with high frequency Pulse Width Modulation so as to obtain high speed control of both dynamic and responsive control. A simple and effective technique of creating an HVDC Light transmission is while combing the converters with a pair of HVDC (AFD, 2010). In principal, the difference between the typical HVDC and HVDC light could have a significant impact on the AC-DC interaction as the VSC allows for more robust support of the AC network besides the power transmission efficiency increase. These features are directly related with the traditional AC system operation. As the HVDC light operates through the DC voltage source including a fast switcher that can turn on and off the existing current, it allows receiving the average value of any voltage in between the two extremes voltages. Additional, characteristics of the HVDC light cables over their HVAC equivalents are related to their lighter weight and other dimension, which leads into a higher power density. Furthermore, the HVDC Light cables have a higher efficiency with decreased transmission losses, they can be laid in deeper waters and on rough bottoms with reduced costs through the ploughing technique and due to that it is supplied in lesser, both in weight and diameter dimensions (Papadopoulos, 2008).

Cables' Technical Features

In general, the structure of the cable is configured in such a way to ascertain reduction of losses, increased efficiency as well as a robust mechanical resistance, a sufficient insulation and magnetic shielding. Different types of cables with diverse technical features may use diverse materials and layout based on the different manufacturers and environmental conditions. Cables use copper in one of the various layers covering the conductor in order to guarantee the physical insulation, impermeability, mechanical strength as well as flexibility and electrical and magnetic protection. While dividing the submarine cables into the two current technologies, it is noticed that the HVDC submarine cables incorporate one primary conductor through which the current is transmitted as well as a pathway to go back either via a second conductor or through the seawater by an anode/cathode. The HVAC cables use three conductors to transmit the current, as aforementioned, which are insulated against any external hazard for the total length of the cable. In terms of cables submarine insulation the principal categories exist (Nexans, 2003; Burnett & Beckman, 2013; Ardelean & Minnebo, 2015) are:

1. Fluid filled (which usually contains lead sheath for water blocking
2. Paper insulated (lapped insulated) cables
3. The new generations, of mass-impregnated paper or XPLE insulations

The latest one demonstrate to be a robust technology which can endure high forces and frequent flexing at the seabed (Asplund et al., 2000; Burnett & Beckman, 2013). Nowadays, XLPE insulations find wide applications in all the synchronous submarine cable installations as presented in Table 1.

Installation and Repair of Submarine Cables

In general, the structure of the cable is configured in such a way to ascertain reduction of losses, increased efficiency as well as a robust mechanical resistance, a sufficient insulation and magnetic shielding.

Installation

Subsea cables installation experiences a number of risks due to the special conditions of the submarine environment such as the unfavorable submarine conditions often encountered, marine environment, transitions between water and land, as well as difficulties to collect geophysical data without overlooking high installation and retrieval stresses on the cable. On top of that, manmade obstacles may be found in the designed submarine cable passageway, such as other cables communication and petroleum pipelines, as well as sewer, water, and gas lines. A way to mitigate the risk could be the joint installation projects, or combing power and communications into the same cable. Other hazards may occur from effluent outfalls, sunken ships and debris, particularly close to harbors, bridges, and other constructions which

Table 1. Materials suitable for different type of cables and their characteristics

Category	Description	Diameter Dimensions	Weight
Fluid filled	• Suitable for high voltages (< 500kV) • No hydraulic limitations • Short distances	110 - 160 mm Conductor: 3000 mm2	40-80 kg/m
Paper insulated	• Suitable for high voltages (< 600kV) • Mass-impregnated cables are the most used	110 - 140 mm Conductor: 2500mm2	30-60 kg/m
The Extruded cables	• Suitable for voltages (< 300kV) • Voltage Source Converters (VSC) • The max transmissible power for VSC with extruded cables is up to 800 MW • However, uneven distribution of charges inside the insulation	90 - 120 m	20-35 kg/m

Data Source: Ardelean & Minnebo, 2015.

could be abandoned and not visible from the surface of the water. Last but not least, disposal areas, either from dredging or dumping of restricted areas for example, naval training or testing areas could process potential risks for placing submarine cables (IEEE, 2005).

The necessary equipment to install submarine cables incorporates special ships and barges which undertake the precise positioning of cables lying on or beneath the seabed, directed by the route survey. Also, in case of shallow waters, where floats could hinder its smooth installation, small boats and divers may be used to support the installation of the cables as determined by the route survey. In contrast, in case of deep water depths remotely Operated Vehicles (ROVs) are incorporated to assist in the cables installation (Burnett & Beckman, 2013).

Cables are usually buried 1 m and in special circumstances up to 10 m under the seabed (Burnett & Beckman, 2013). This is considered a reasonable range of depths to protect the cables from trawl fishing, anchoring and other activities. As a rule, in case of compound cables that are placed in the same position, they are buried in a configuration to leave a certain distance away from each other, to allow for easy maintenance. During the installation process, there a number of challenges might be encountered related to possible inappropriate qualities in the seabed texture, in the corridor buried and turbid water. Nonetheless, the level of the impact is directly related to the burial method, seabed type and wave/current action. Usually, cabling design studies assess the various forms of disturbance from and to the cables. They usually indicate that the impact from fishing and other activities are short-term in the sense of months, where waves/currents are energetic, however usually in a larger extent in deeper and less turbulent waters (IEEE, 2005; Burnett & Beckman, 2013; Ardelean & Minnebo, 2015).

The stage of the cables burying is usually the one that causes the most significant environmental impact mainly on the flora and the fauna. While the installation process is ended and the cable is secured and fastened the sea life generally succeeds to recuperate in a short period of time (Ardelean & Minnebo, 2015). The environmental impact assessment study beforehand the installation, covers the different aspects of transmission installation, operation and maintenance including the indicative burying techniques and protection measures. Usually, the sub-sea living creatures will move from the area that the cables are installed; however the main effect is noticed on motionless species that use the seabed to develop and live. A number of studies exploring this subject suggest staying away from areas where scarce or hard recovering species exist while diminishing the cable movements on the seabed both during installation and operation. However, they mention that usually there is a zero impact over the benthonic species and dynamics of sediment (Andrulewicz et al., 2003) and a 55% improvement in the submarine fauna one year following the completion of the cable's installation, while after two years this amount reaches approximately 85% (Dunham et al., 2015).

Repair

Cables usually operate in a reliable way although they are threatened by a number of different hazards usually caused by mechanical destruction from human activities. Possible failures may arise from: physical damage caused by human activities, forces of nature, such as seismic activity, electrical insulation breakdown, hydraulic failure (for fluid-filled cables). With regards to the human activity, potential hazards could be ships' anchors and tug boat lines, beached marine equipment, dock and bridge maintenance, redging, dumped debris, fishing activity, shellfish harvesting, aquatic farming, pile driving, horizontal directional drilling, other cable or pipe laying operations. Finally, maintenance work on adjacent cables or pipelines, contamination of the seabed with chemicals, toxins, and heavy metals could reveal potential risk for submarines cable operation (IEEE, 2005).

The repair of cables including also replacement or calibrating is usually a complex process which might take days or weeks to be completed dependent also on the weather conditions and the extent of the damage. A possible damage could affect the supply of whole regions and usually constitutes a major event which requires immediate intervention. During the repair of a damaged cable are required specialist ships and cable jointing experts (divers) while using cameras to replace the damaged section with new cable. An appropriate way of dealing with these kinds of incidents before proceeding to the actual repair could be to use the documentation and search for the evidence which caused the failure. This could be performed by locating electronic techniques which could assist on identifying where the damage is and other geophysical features of the surrounding area, to prepare for the repairing process. In case the cables belong to the SCFF category, dielectric fluid leaks could be identified by checking the metering for malfunctions and calibration, to test the pressure-drop and perform air-surveillance (IEEE, 2005; Burnett & Beckman, 2013) .

INTERCONNECTIONS, PRACTICES AND EXPERIENCES WORLDWIDE

Interconnection between islands or between mainland and one (or more) isolated islands are deployed all over the world. Let's take you in a journey of events across the world in time. We mapped interconnections projects (HVDC, HVAC and Future projects) (Figure 1) with some unique characteristics and discussed key features in parallel with the progress made in terms of materials and interconnections technical power. The key steps in history of interconnections of islands with mainland or between islands together with key milestones projects are discussed bellow (for summary see Table 2, Appendix 1).

As aforementioned, the first island ever interconnected was the Gotland Island in 1954. Currently, the longest island interconnection in the world is the one between Sardinia and Italy equal to 420 km which is in the same order of magnitude as the largest sub-sea inter-connector, the NorNed cable between Norway and the Netherlands equal to 580km, with a capacity of 700 MW (Ardelean & Minnebo, 2015). According to the last technological indications the very latest cable technology has the potential capability of reaching up to 1,500 km. The longest planned interconnectors in the world will connect Iceland with UK and will have a length of 1170 km, expected to be commissioned by 2022 followed by the Euroasia interconnector which will interconnect through 1000 km of submarine cables, Israel, Cyprus, Crete island and the Greek mainland (EuroAsia interconnector, 2013).

The majority of islands interconnection projects have been recorded in Asia (Table 4), although the longest and more concrete projects are found in Europe (Table 3). In Asia, the islands are usually located close to the shore 10-55 km or in island complexes such as Indonesia and Philippines where the enhancement of the national grids through interconnections with then neighboring islands is relatively easy achieved through short HVAC links. In Europe, longer projects are observed exceeding the 400 km mainly in the Mediterranean basin, while the new HVDC interconnections are expected to reach even longer lengths. In North America, only a few island interconnections have been implemented in close distances from the shore (Table 5). Currently, the second longest submarine island interconnection, after Sardinia-Italy link is the Basslink connecting Tasmania with Australia (Table 6). The main driver to interconnect islands has been principally the requirement to access cheap energy sources usually located in the mainland,

Figure 1. Map of islands interconnections

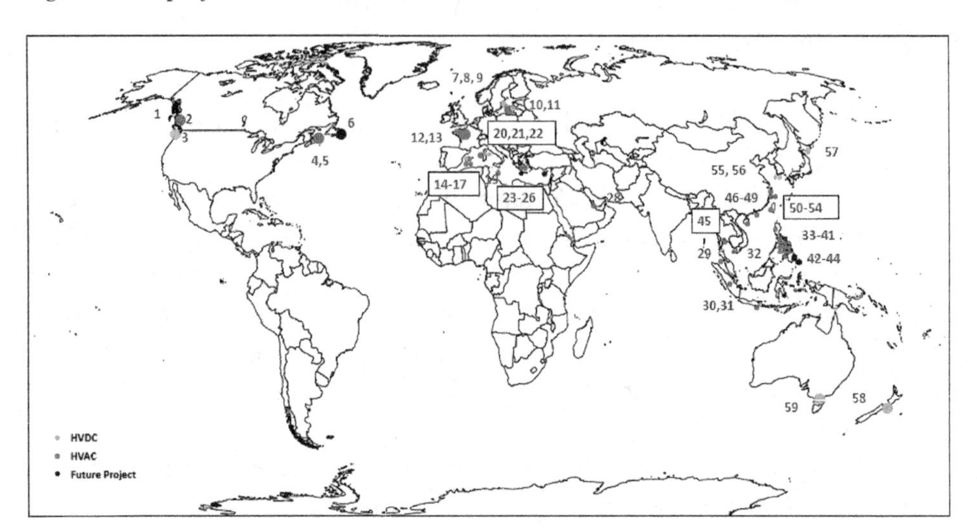

and secondly the enhancement of the local power networks in order to increase renewable energy sources on the isolated regions, which usually possess a high renewable energy potential. With the exception of St. George Island, which is an uninhabited island that hostess a wind farm as well as the Labrador Island interconnection project where the main driver is a large scale wind farm of 824 MW that had to be interconnected with the wider Canadian mainland, the rest of the projects included in this study (Figure 1) were implemented driven by the combination of the two aforementioned requirements.

Initially, submarine power cables were used to connect shore-based power grids across bays, rivers, estuaries, canals, etc. Nowadays, the need to mitigate energy isolation as well as the increasing energy demand in isolated regions, led oversea regions to connect, by laying down cables which link power grids at different regional levels: islands, countries, continents interconnections. In addition, the decentralization of energy generation sources resulted into the significant progress of submarine cabling technology being used in offshore installations, e.g. oil/gas platforms, offshore wind farms, wave and tidal energy installations and ocean science observatories. The first submarine cable was installed in 1811 across the Isar River in Bavaria, Germany insulated with natural rubber, while later in 1924 lead extrusion was introduced as a water barrier in the submarine cables. By 1937, the first synthetic insulation cable-butyl rubber was presented and by 1952 the cables started using oil-filled insulation.

European Islands started in force. In 1954 (Appendix 2, Table 3), the world's first commercial transmission link was connecting Gotland and mainland Sweden. Its voltage was 100 kV and the capacity 20 MW. The length was 96 km long cable. The event changed electricity supply on the isle of Gotland. This project has been also the first submarine transmission incorporating HVDC technology featuring a 20 MW, 150 kV and 96 km line. The Gotland island interconnection proceeded to several upgrades within the last decades (ABB, 2014). In 1985 due to increased demand and concern about supply safety, another HVDC link to Gotland was build, the Gotland 3, which usually works with Gotland 2 to form a bipolar link but can also work independently, with a total transmission capacity of 260MW. Later in 1986 Gotland 3 was built and Gotland 1 was taken out of service. In 2017, ABB installed its MACH control and protection system, allowing high degree of integration capability to handle control and protection functions for 30 years at least (ABB, 2018). The upgrades were driven by the evolution of materials and cables.

Very soon after the first interconnections, in 1962 the market welcomed the first ethylene-propylene rubber insulation which was mainly replaced in 1990 by linked polyethylene (XLPE) insulation initially introduced in 1973 (Burnett & Beckman, 2013). Interconnection among islands or between islands and the mainland supports isolated power networks located on the islands, while transmitting electricity from larger networks or while creating and enhancing regional grids among the islands. These interconnections are usually financially supported by the European Union or national governments worldwide in order to provide electricity connection to new customers for the first time while eliminating the use of expensive and emission pollutant local generation relying on heavy fuel oil and diesel fuel. The largest part of the islands distribution or transmission subsea circuits belong to the AC technology category at medium or high voltages up to 150 kV (Subsea Cables UK, 2013). Islands interconnections depend mainly on single radial circuits whilst frequently multiple circuits or interconnected networks are required to meet electricity demand while improving and enhancing reliability and the security of supply in the local networks (Ardelean & Minnebo, 2015).

In 1965, the HVDC interconnection between Italy-Corsica-Sardinia also named SACOI 2, has some unique characteristics, namely it has more than two converter station as part of the a single HVDC system. Is one of the only two Multi-terminal HVDC systems in operation in the world. Corsica is the smaller island to the north, while Sardinia is the second largest island in the Mediterranean, situated due west of mainland Italy. Currently this has been closed to the end of the useful lifetime and not available anymore after 2023. A tri-terminal HVDC link will link Italy mainland, Corsica and Sardinia (Project 299- SAC013). However, the SACOI 2 link is getting old and there are uncertainties on their reasonable residual life (Ardito et al., 2004). Later in 1972 the world's first submarine XLPE cable connecting the Swedish mainland to the Finish island Åland was installed. The cable link consist of three cable 55 km each, for 84 kV.

Mallorca and Menorca are interconnected in 1981. Only later in 2007, a 69 kV-cable with a capacity of 50 MW is commission in order to interconnection the power-grid of Formentera island with that of Ibiza. Also a 132 kV Line installed between Majorca and Minorca in the Balearic isles. In 2016 the two electricity subsystems (Mallorca-Menorca and Ibiza-Formentera were connected by the Majorca-Ibiza double electricity link. At the time of its commissioning, this interconnection link is the longest and deepest submarine link in alternating current of its kind in the world, as the submarine cable rests on the seafloor at depths of up to 800 meteres.

In 1985, Jersey Island and France was interconnected for first time. Since then the link enabled 50% of the island of Jersey's electrical needs to be satisfied. A second link was commissioned in 2000, bringing its electrical power capacity up to 135 MW to meet variations in demand throughout the year, as well as consumption peaks, with a wide enough margin. In parallel, an undersea cable was installed between Jersey and Guernsey, making the second Jersey-France interconnection doubly useful, as it would supply both islands (RTE Electricity Grid System Manager, 2001).

The year of 2009 for islands in EU, the DIRECTIVE 2009/72/EC (European Parliament, Council of the European Union) was a key official document, concerning common rules for the internal market in electricity & incorporating isolated systems forming electricity islands. Later The European Economic and Social Committee publishes its Opinion "Getting EU energy islands connected: growth, competitiveness, solidarity and sustainability in the EU internal energy market" towards interconnections, sustainability and affordability in EU energy islands (Eur-Lex)

Asia (Appendix 2, Table 4)

In Asia many interconnections took place during time. Here are some examples.

- **1989:** An early subsea KLPE cable installation was a 132 kV cable link in Hong Kong (Wan et al., 1990)

- **1990:** Negros-Panay interconnection in Philippines (Features: 68 kV,18 km)
- **1992:** Upgrade of the Sardinia-Corsica Interconnection (Features: double-pole, 300 MW)
- **1993:** Cebu-Negros interconnection in Philippines took place (Features: 138 kV,18 km, 90MW)
- **1994:** Java-Madura-Bali Interconnection completed through network (500 kV) and SUTT (150 kV and 70 kV) in Indonesia
- **1996:** The HVDC Haenam–Cheju between the Korean Peninsula and the island of Jeju in South Korea was installed. (Features: 101 km, HVDC bi-polar, 300MW)
- **1997:** Leyte-Cebu in Philippines Features: 230 kV, 17 km, 185 MW was put into service)
- **1998:** ASEAN Power Grid program (APG) identified and initiated in order to establish an Integrated ASEAN Power System between several countries comprising of islands: Malaysia, Thailand, Philippines, Cambodia, Myanmar, Vietnam and others Leyte-Luzon in Philippines Interconnection Features: 440MW, 23km 350kV DC
- **2000:** The longest 500- kV cable in the world, the Shin-Toyosu line, was installed in Japan by the Tokyo Electric Power Company. This double circuit line has four 300-MVar shunt reactors for the compensation of the large charging capacity (Ametani et al., 2016)
- **2001:** The Philippines enacted the Republic Act 9136 entitled as "Electric Power Industry Reform Act" in order to ensure and accelerate the total electrification of the country, security and affordability to the islanding region
- **2004:** The construction of a 2-staged interconnection project in Philippines was completed. Stage I, included the construction of a submarine cable between Leyte and Bohol by 69-kV lines. In Stage II, related transmission lines and substation facilities were constructed in 138 kV.
- **2005:** Cebu-Mactan Interconnection in Philippines (Features: 200 MW, 138 kV, 8.5 km) Enhancement of Leyte-Cebu in Philippines (Features: 230 kV, 17 km, 185 MW)
- **2006:** Panay-Boracay interconnection in Philippines (Features: 69kV, 1.6 km)
- **2007:** Enhancement of Cebu-Negros interconnection in Philippines (Features: 138 kV,18 km, 90MW)

- **2008:** A 33kV submarine cable to Koh Siboya, Koh Pu, Koh Phi Phi, Railey Bay, and Koh Lanta islands was submerged. (Features: total submarine 62.8 km, in two lengths, of XLPE insulated, lead sheathed, copper conductor 3-core 33 kV submarine cable, with an integrated fibre optic element)
- **2010:** ASEAN Interconnection Master Plan Study (AIMS) II agreed & published, prioritizing an number of future interconnections in the area of the South Eastern Asian Nations. The interconnection of Al Jasra (Bahrain), Ras Qurray (Saudi Arabia) sub stations, and Umman Na'san Island takes place (Features: 120km, HVAC, oil-filed, 400kV)
- **2011:** ShangXiaChuan Island Interconnection with China (Features: 64/110kV, 30.293km).
- The Taiwan Power Company commenced a project launched to manufacture and lay six submarine power cables between the island of Taiwan and the Penghu Islands, approximately 60 km off the southwest coast of the island of Taiwan.
- **2012:** SiJiao island interconnection with the China mainland (Features: 64/110kV, 34.4kM)
- SiJiao island interconnection with the China mainland (Features: 64/110kV, 34.4kM)
- ZhouShan Island Interconnection with China is completed (Features: 2009-2012) project≈103 km.
- **2014:** The proposed interconnection of Borneo island (Malaysia) with Mindanao (Philippines) has been proposed for funding in the Asian Development Bank
- **2015:** Phu Quoc Island is interconnected to the national power grid in Vietnam (56 km of 3x630 mm^2 110 kV copper, single wire armoured XLPE insulated submarine cable)
- The new submarine cable link from Guangdong power grid to Hainan island has been awarded (HVAC, 500 kV, single core, self-contained fluid filled cables, 32 600 MW capacity)
- **2016:** The Jindo–Jeju, submarine cable was commissioned to interconnect: South Korea (Features: 105 km HVDC, 400 MW, 250 kV)

North America Europe (Appendix 2, Table 5)

- **1968:** The Submarine cable connection between the Canadian mainland and Vancouver Island went into service. Later in 1977 the HVDC Vancouver Island link was supplemented by installing a second pole. In 1984 the world's first long 500-/400-kV cable was installed in Canada by BC Hydro. This is a double-circuit line that connect Vancouver island to mainland Canada through Texada Island (Ametani et al., 2016). In 2014 The Muskrat Falls Project is agreed including a hydro project of 824MW and the interconnection of Labrador (Canada) to Strait of Belle isle-Newfoundland island and Nova Scotia Island. Two year later, in 2016 The Prince Edward Island's electricity interconnection was completed (2*180MW) with Northumberland Strait and New Brunswick (Canada). The Project required to address the growing electricity demand & to deliver reliable, long-term energy for Islanders through energy imports.
- **2003:** The utility GRENLEC has initiated a comprehensive Interconnection policy. For distributed generation bind in the electrical grid for the island of Grenada in the Caribbean Islanding Region
- **2004:** A Task Force has been established, comprising several Caribbean Islands to develop recommendations for a Regional Energy Policy, which would address several issues of security, pricing and the impact on relative competitiveness in the CARICOM Single Market and Economy. Later in 2010 the World Bank establishes a strategy on Caribbean Regional Electricity Generation, Interconnection, and Fuels Supply. The United States Agency for International Development (USAID) publishes the Maldives Submarine Cable Interconnection Pre-feasibility Study, and proposed a number of interconnection projects in the Maldives islands complex Virgin Islands Water and Power Authority proposed a transmission project that interconnects the electrical power systems on the islands of: Puerto Rico (PREPA) St. Thomas and St. Croix (USVI) the British Virgin Islands (BVI)
- **2012:** The World Bank assesses Regional Electricity Supply Options & Electricity market interconnections for the Caribbean Islands St. Kitts and Nevis (Caribbean Islands) interconnection with Puerto Rico study is launched in order to transfer large geothermal capacity located in the two islands

- **2013:** An amended CARICOM Energy Policy was approved alongside and the Caribbean Sustainable Energy Roadmap and Strategy for interconnections

Australia (Appendix 2, Table 6)

- **1961:** Agreement to connect in the in New Zealand the North where was high population density with the South which had high potential for hydroelectric power
- **2005:** The commissioning of the Basslink between Australia and Tasmania was put into operation (Features: 370monopolar, HVDC, 400 kV. The nominal rating of the link is 500 MW although it is capable of transmitting 630 MW). However, in December 2015 there was a failure of the Basslink cable due to a tear in the interconnector. In December 2017, the ABC reported that following the investigation into the cable's performance, the damage in the cable may have happen when running above continuous rating, through overheating, due to inadequate design and operation. This determine Tasmania government to outline how the state would become 100% renewable by 2022.

DISCUSSION AND FUTURE RESEACH DIRECTIONS

As it can be seen since the islands ‘ interconnections came into commercial operation, many isolated islands' electrical systems came to an end, bringing security of supply, favouring a greater integration of renewable energy, reduce the electricity cost in the island and improve environmental conditions in the islands. For example, to date all the Ionan Islands have been interconnected to the High Voltage system. However, putting in place a cable may seem easily implementable from technical perspective, but may require serious decision-making and political support at local, regional and inter-regional level. Usually islands located close to mainland are already interconnected. Those who are located in remote areas are more difficult to connect due to high cost of large distance cables and maintenance. However, more progress could be made in interconnected neighbouring islands. Literature provides case studies analyzing the value of island interconnections (e.g. interconnection between between Malta and Sicily (Ries et al., 2016), the value of interconnection links in remote island power systems: The Spanish Canary and Balearic archipelago cases (Lobato et al., 2017), the effect of interconnecting Greek islands to the mainland system via electric submarine cables (Georgiou et al., 2011). The latest concluded that development of interconnections represents the optimum solution succeeding in the curtailment of total electricity generation cost. Furthermore interconnected them with the mainland, will provide economic and environmental benefits.

General benefits of grid interconnections include: improve reliability and pooling reserves, reduce investment in generating capacity, improve load factor and increase load diversity, diversity of generation mix and supply security. However, problems could occur in terms of market integration. For example, in the case of the interconnection cable from Malta to Sicily, which belongs to two EU member states. Until 2015, Malta relied on imported fossil fuels (oil), having one vertically integrated corporation serving consumers (Enemalta, 2007). A recent study (Ries et al., 2016) showed that the newly installed interconnector between Malta and Sicily does not necessarily lower electricity prices for Malta's consumers and that cable impacts depend on the installed generation capacity, oil price, and market design. Despite this, an increasing utilization of renewables is a great opportunity to reduce environmental impacts due to use of fossil fuels and provide a better and healthier life for further generation. Careful consideration should be given to the integration of fluctuating renewables in small and weak islands grids, which could be a problem (Tsuchica, 2014). To tackle such problem, energy storage could be the solution to decouple electricity supply and demand thus providing flexibility and the ability to provide system services, such as frequency regulation and voltage control, spinning reserve (Rodrigues et al., 2014). Some studies have analysed the techno-economic aspects of applying storage systems on islands (Parissis et al., 2011; Camus & Farias, 2012; Hlusiak et al., 2012; Pina et al., 2012; Correia et al., 2014; Lopes et al., 2016; Child et al. 2017). According to Cozzolino et al. (2016) it is possible to replace fossil electricity generation up to 100% if we combine renewable energies and storage systems in hybrid power plant approaches. Furthermore, the literature provides us with some comprehensive reviews on the utilization of renewable energies on islands (Notton, 2015; Kuang et al., 2016).

The smooth integration of renewable energy systems into the electricity system will only be possible if the system becomes much more flexible. In small islands with a good understanding of the energy demand and customers profiles, it is easier to manage it than to achieve larger interconnected networks. If we introduce in the system larger electrical loads and electric vehicles and charging points, then the network operator could switch loads on and off to finally produce flat load curve via use of time of day tariffs. This will bring a number of benefits, removes system peak, increases asset utilization and lower capital expenditure.

Considerable opportunities exist in islands to embrace innovative solutions to improve energy resource use, manage the whole energy system. More studies are needed to test different options. Such solution could be integration of heat pumps technologies for heating and cooling. More studies are needed to understand the benefits of interconnections vs storage in terms of economics and security of supply.

CONCLUSION

From this chapter, a number of key conclusions can be drawn.

There is much progress done in terms of submarine power transmission technology which have been already proven that can help improve quality of supply, facilitate the development of local renewable energy sources, and help reduce environmental impact. Furthermore, interconnecting small islands could help reduce intermittency and variability, and increase resilience of the grid.

HVDC Light Cables are well suited to be installed under severe conditions, both as an underground land cable and as a submarine cable. The polymeric cables has proven to be suited to work in difficult areas, such as seismic areas, areas of extreme water depths and so on

In terms of cost, this vary from case to case; however in some cases it proved to be cost effective solution, in others an expensive solution.

More research is needed to understand if there any other effects (e.g. noise pollution, impact of operation of subsea cables on fauna, if pose any risks to marine fauna; impact on the biological environment, if heat dissipation could have ecological consequences and so on). Also a better understanding on impact on consequences after end of life of cables, what is the effect of the process of removing the cables or contamination effect if are left, the risk they pose for harming marine fauna.

REFERENCES

ABB. (2006). HVDC Light ® Cables, *Submarine and land power cables.* Retrieved from: https://library.e.abb.com/public/564b3711c060164dc1257b0c00552e50/HVDC%20Light%20power%20cables.pdf

ABB. (2008). *ABB 'world-firsts' in high voltage cables.* Retrieved from: http://www.abb.com/cawp/seitp202/48b3feb9af18e957c125737800474543.aspx

ABB. (2014). *ABB Review Special Report 60 years of HVDC.* ABB Group R&D Technology. Retrieved from: https://library.e.abb.com/public/aff841e25d8986b5c1257d380045703f/140818%20ABB%20SR%2060%20years%20of%20HVDC_72dpi.pdf

AFD. (2010). *Implementing Large-Scale Energy Efficiency Programs in Existing Buildings in China.* AFD.

Andrulewicz, E., Napierska, D., & Otremba, Z. (2003). The environmental effects of the installation and functioning of the submarine SwePol Link HVDC transmission line: A case study of the Polish Marine Area of the Baltic Sea. *Journal of Sea Research, 49*(4), 337–345. doi:10.1016/S1385-1101(03)00020-0

Ardelean, M., & Minnebo, P. (2015). *HVDC Submarine Power Cables in the World.* EUR 27527 EN. doi: 10.2790/95735

Asian Development Bank (ADB). (2013). *Proposed Loans Republic of Indonesia: Java-Bali 500-Kilovolt Power Transmission Crossing Project.* Retrieved from: https://www.adb.org/sites/default/files/project-document/79587/42362-013-rrp.pdf

Asplund, G., Eriksson, K., & Tollerz, O. (2000). *Land and sea Cable interconnections with HVDC Light.* CEPSI 2000 Conference, Manila, Philippines.

Baltic Transport Journal. (2013). *HVDC between Finland and the Aland Islands.* Retrieved from: http://baltictransportjournal.com/search/finland/hvdc-between-finland-and-the-aland-islands,37.html

Bresesti, P., Kling, W. L., Hendriks, R. L., & Vailati, R. (2007). HVDC Connection of Offshore Wind Farms to the Transmission System. *IEEE Transactions on Energy Conversion*, *22*(1), 37–43. doi:10.1109/TEC.2006.889624

Brito, M. C., Lobato, K., Nunes, P., & Serra, F. (2014). Sustainable Energy Systems in an Imaginary Island. *Renewable & Sustainable Energy Reviews*, *37*, 229–242. doi:10.1016/j.rser.2014.05.008

Burnett, D. R., Beckman, R., & Davenport, T. M. (2013). *Submarine Cables. In The Handbook of Law and Policy* (p. 488). Martinus Nijhoff Publishers; doi:10.1163/9789004260337

Camus, C., & Farias, T. (2012). The electric vehicles as a mean to reduce CO2 emissions and energy costs in isolated regions. The São Miguel (Azores) case study. *Energy Policy*, *43*, 153–165. doi:10.1016/j.enpol.2011.12.046

Catalão, J. P. S. (2014). Energy storage systems supporting increased penetration of renewables in islanded systems. *Energy*, *75*, 265–280. doi:10.1016/j.energy.2014.07.072

CBC News. (2016). *$80M loan approved for electric cable project*. Retrieved from: http://www.cbc.ca/news/canada/prince-edward-island/pei-cable-loan-power-electricity-1.3476000

Child, M., Nordling, A., & Breyer, C. (2017). Scenarios for a sustainable energy system in the Åland Islands in 2030. *Energy Conversion and Management*, *137*, 49–60. doi:10.1016/j.enconman.2017.01.039

Cirio, D. (2016). *DEMO 3 Upgrading of Multi-Terminal Interconnectors, RSE "Upgrading Europe's Grid to new high-capacity technologies"*. Best paths dissemination workshop, Berlin, Germany. Retrieved from: http://bestpaths-project.eu/contents/publications/berlin_de_5_demo3_-diego-crio.pdf

4. Coffshore. (n.d.). *Normandie 1 (N1) Interconnector*. Retrieved from: http://www.4coffshore.com/windfarms/interconnector-normandie-1-(n1)-icid78.html

Correia, P. F., Ferreira de Jesus, J. M., & Lemos, J. M. (2014). Sizing of a pumped storage power plant in S. Miguel, Azores, using stochastic optimization. *Electric Power Systems Research*, *112*, 20–26. doi:10.1016/j.epsr.2014.02.025

Cozzolino, R., Tribioli, L., & Bella, G. (2016). Power management of a hybrid renewable system for artificial islands: A case study. *Energy*, *106*, 774–789. doi:10.1016/j.energy.2015.12.118

Daniel, J. P., Liu, S., Ibanez, E., Pennock, K., Reed, G., & Hanes, S. (2014). National Offshore Wind Energy Grid Interconnection Study. *DOE Award*, *33*(No. EE-0005365), 243.

Diagne, M., David, M., Lauret, P., Boland, J., & Schmutz, N. (2013). Review of Solar Irradiance Forecasting Methods and a Proposition for Small-Scale Insular Grids. *Renewable & Sustainable Energy Reviews*, *27*, 65–76. doi:10.1016/j.rser.2013.06.042

Dunham, A., Pegg, J.R., Carolsfeld, W., Davies, S., Murfitt, I., & Boutillier, J. (2015). Effects of submarine power transmission cables on a glass sponge reef and associated megafaunal community. *Marine Environmental Research*, *107*, 50–60. doi: 10.1016/j.marenvres.2015.04.003

Eur-Lex Access to European Union Law. (n.d.). *Opinion of the European Economic and Social Committee on 'Getting EU energy islands connected: growth, competitiveness, solidarity and sustainability in the EU internal energy market'*. Retrieved from https://eur-lex.europa.eu/legal-content/EN/TXT/?uri=CELEX%3A52012AE1696

EuroAsia Interconnector. (2013). *A European Union Project of Common Interest*. Retrieved from: http://www.euroasia-interconnector.com/

Gaji, Z., Hillstrom, B., & Mekic, F. (2003). HV Shunt Reactor Secrets for protection engineers. *30th Western Protective Relaying Conference*.

Georgiou, P., Mavrotas, G., & Diakoulaki, D. (2011). The effect of islands' interconnection to the mainland system on the development of renewable energy sources in the Greek power sector. *Renewable & Sustainable Energy Reviews*, *15*(6), 2607–2620. doi:10.1016/j.rser.2011.03.007

Hlusiak, M., Arnhold, O., & Breyer, C. (2012). *Optimising a Renewables Based Island Grid and Integrating a Battery Electric Vehicles Concept on the Example of Graciosa Island, Azores Archipelago*. 6th European Conference on PV-Hybrids and Mini-Grids, Chambéry.

HVDC Sumantra-Java. (2016). *HVDC Project*. Retrieved from: http://hvdcsumatrajava.com/

IEEE Power Engineering Society. (2005). *IEEE Guide for the Planning*. Design, Installation, and Repair of Submarine Power Cable Systems; doi:10.1109/IEEESTD.2005.95937

East Java Electric Power Transmission and Distribution Network Project (IV). (2002). Retrieved from: https://www.jica.go.jp/english/our_work/evaluation/oda_loan/post/2002/pdf/032_full.pdf

Jovcic, D., & Strachan, N. (2009). Offshore wind farm with centralised power conversion and DC interconnection. *IET Generation, Transmission & Distribution*, *3*(6), 586–595. doi:10.1049/iet-gtd.2008.0372

Kuang, Y., Zhang, Y., Zhou, B., Li, C., Cao, Y., Li, L., & Zeng, L. (2016). A review of renewable energy utilization in islands. *Renewable & Sustainable Energy Reviews*, *59*, 504–513. doi:10.1016/j.rser.2016.01.014

Lobato, E., Sigrist, L., & Rouco, L. (2017). Value of electric interconnection links in remote island power systems: The Spanish Canary and Balearic archipelago cases. *Electrical Power and Energy Systems*, *91*, 192–200. doi:10.1016/j.ijepes.2017.03.014

Nai, C. C. (2015). *Power Interconnections in the Greater Mekong Subregion*. Sustainable Hydropower and Regional Cooperation in Myanmar Nay Pyi Taw.

Nanou, S., Papathanassiou, S., & Papadopoulos, M. (2014). HV Transmission Technologies for the Interconnection of the Aegean Sea Islands and Offshore Wind Farms. *Proc. MedPower 2014*. 10.1049/cp.2014.1674

Nexans. (2003). *Submarine Power Cables*. Retrieved from: http://www.nexans.com/Germany/group/doc/en/NEX_Submarine_neu.pdf

Nexans. (2005). *Nexans wins a 171 million USD order to supply extra-high-voltage submarine link between Saudi Arabia and Bahrain*. Retrieved from: https://www.nexans.com/upload/objects/20051121/Saudi_Bahrain_GB.pdf

Nexant. (2010). *Caribbean Regional Electricity Generation, Interconnection, and Fuels Supply Strategy Final Report*. Retrieved from: http://documents.worldbank.org/curated/en/440751468238476576/pdf/594850Final0Report.pdf

NordPoolspot. (2017). *Principles for determinig the transfer capacities in the Nordic Power Market.* Retrieved from: https://www.nordpoolspot.com/globalassets/download-center/tso/principles-for-determining-the-transfer-capacities.pdf

Notton, G. (2015). Importance of islands in renewable energy production and storage: The situation of the French islands. *Renewable & Sustainable Energy Reviews*, *47*, 260–269. doi:10.1016/j.rser.2015.03.053

Orient Cable. (2011). *Installation.* Retrieved from: http://www.orient-cables.com

Papadopoulos, C., Papageorgiou, P., Stendius, L., Åström, J., Hyttinen, M., & Johansson, S. (2004). *Interconnection of Greek islands with dispersed generation via HVDC Light technology.* Retrieved from https://library.e.abb.com/public/e9c3ec16c33e7937c1256fda004c8ccc/Cigre%20Paper_16.11.2004.pdf

Parissis, O. S., Zoulias, E., Stamatakis, E., Sioulas, K., Alves, L., Martins, R., ... Zervos, A. (2011). Integration of wind and hydrogen technologies in the power system of Corvo island, Azores: A cost-benefit analysis. *International Journal of Hydrogen Energy*, *36*(13), 8143–8151.

Paternò, G., Madonia, A., Ippolito, M. G., Massaro, F., Favuzza, S., & Cassaro, C. (2016). Analysis of the new submarine interconnection system between Italy and Malta : simulation of transmission network operation. In *2016 IEEE 16th International Conference on Environment and Electrical Engineering.* Florence, Italy: IEEE Xplore. 10.1109/EEEIC.2016.7555467

Pina, A., Silva, C., & Ferrão, P. (2012). The impact of demand side management strategies in the penetration of renewable electricity. *Energy*, *41*(1), 128–137.

Prysmian Group. (2015). *Prysmian, first-ever submarine power cable award in China.* Retrieved from: https://www.prysmiangroup.com/en/en_2015-PR_Hainan.html

Red Electrica de Espana. (2014). *Red Eléctrica begins the cable laying works for the submarine cable between Ibiza and Majorca.* Retrieved from: http://www.ree.es/en/press-office/press-release/2014/12/red-electrica-begins-cable-laying-works-submarine-cable-between-majorca-ibiza

Red Electrica de Espana. (n.d.). *Romulo Project*. Retrieved from: http://www.ree.es/en/activities/unique-projects/romulo-project

Reidy, A., & Watson, R. (2005). Comparison of VSC based HVDC and HVAC interconnections to a large offshore wind farm. *IEEE Power Engineering Society General Meeting*, (1), 1–8.

Ries, J., Gaudard, L., & Romerio, F. (2016). Interconnecting an isolated electricity system to the European market: The case of Malta. *Utilities Policy*, *40*, 1–14. doi:10.1016/j.jup.2016.03.001

Rodrigues, E. M. G., Godina, R., Santos, S. F., Bizuayehu, A. W., Contreras, J., Setas Lopes, A., ... Ferreira de Jesus, J. M. (2016). Contributions to the preliminary assessment of a Water Pumped Storage System in Terceira Island (Azores). *Journal of Energy Storage*, *6*, 59–69. doi:10.1016/j.est.2016.01.009

RTE Electricity Grid System Manager. (2001). *Presentation of the second Jersey-France Interconnection*. Retrieved from: http://clients.rte-france.com/htm/an/journalistes/telecharge/dossiers/jersey_france_an.pdf

Ruddy, J., Meere, R., & Donnell, T. O. (2015). Low Frequency AC Transmission as an alternative to VSC-HVDC for grid interconnection of offshore wind. *IEEE Powertech Conference*. 10.1109/PTC.2015.7232420

Stantec. (2017). *Marine Cable Trenching Addendum: PEI – NB Cable Interconnection Upgrade Project*. Retrieved from: https://www.princeedwardisland.ca/sites/default/files/publications/cable_interconnection_upgrade_addendum_web.pdf

Subsea Cables, U. K. (2013). *Submarine Power Cables Ensuring the lights stay on!* Retrieved from: http://www.escaeu.org/articles/submarine-power-cables/

T&D World. (2014). *Completion of Operations for SE Asia's Longest Undersea Cable System*. Retrieved from: http://www.tdworld.com/underground-tampd/completion-operations-se-asia-s-longest-undersea-cable-system

Tsuchida, T. B. (2014). *Renewables Integration on Islands A2. In Renewable Energy Integration* (pp. 295–305). Boston: Academic Press.

Viohalco. (2015). *Fulgor, Hellenic Cables Group: Interconnection Project for transmitting power from the wind farm located in Agios Georgios Island*. Retrieved from: http://viohalco.com/Article/14/el/

Wu, Y., Chang, G. W., & Chiang, M. H. (2013). Black-Start Dynamic Simulation for Taiwan-Penghu Interconnection System, Published in Automatic Control Conference (CACS). *2013 CACS International.* doi: 10.1109/CACS.2013.6734173

Zafeiratou, E., & Spataru, C. (2016). Transforming the Greek Cycladic islands into a wind energy hub. *Proceedings of the Institution of Civil Engineers-Engineering Sustainability*, *170*(2), 113-129.

Zafeiratou, E., Spataru, C., & Bleischwitz, R. (2016). Wind offshore energy in the Northern Aegean Sea islanding region. In *EEEIC 2016 - International Conference on Environment and Electrical Engineering*. Florence, Italy: IEEE Explore. 10.1109/EEEIC.2016.7555518

APPENDIX 1: HISTORICAL MILESTONES OF ISLANDS INTERCONNECTIONS

Table 2. Islands: Historical key milestones of interconnections

Year	Asia	Europe	North America	Australia
1954		**The world's first commercial HVDC transmission link connected Gotland and the Sweden mainland (Features: 20 MW, 150 kV,96 km)**		
1961			**The Submarine cable connection between the Canadian mainland and Vancouver Island is completed.**	Agreement to connect in the in New Zealand the North where was high population density with the South which had high potential for hydroelectric power
1965		Corsica & Sardinia Interconnection to Italy (Features: AC, monopole, 200kV, 200MW)		
1972		**Aland Island in Sweden becomes interconnected with the mainland (84kV XLPE cable)**		
1983		**Enhancement of the Gotland-Sweden Cable to 130 MW the first in the world to feature a fully redundant digital control and protection system and gas insulated switchgear (GIS)**		
1985		**Jersey Island part of the Channel islands (UK) was interconnected with France (paper insulation technology)**		
1987		**The first cable of 1954 was replaced in Gotland-Sweden Interconnection. A bipolar link has been formed. The total operational transmission capacity is 260 MW, but it has a max capacity of 320 MW.**		

continued on following page

Table 2. Continued

Year	Asia	Europe	North America	Australia
1990	Negros-Panay interconnection in Philippines (Features: 68 kV,18 km)			
1992	Upgrade of the Sardinia-Corsica Interconnection (Features: double-pole, 300 MW)			
1993	Cebu-Negros interconnection in Philippines took place (Features: 138 kV,18 km, 90MW)			
1994	*Java-Madura-Bali Interconnection completed through network (500 kV) and SUTT (150 kV and 70 kV) in Indonesia*			
1996	**The HVDC Haenam–Cheju between the Korean Peninsula and the island of Jeju in South Korea was installed. (Features: 101 km, HVDC bi-polar, 300MW)**			
1997	Leyte-Cebu in Philippines Features: 230 kV, 17 km, 185 MW was put into service)			
1998	ASEAN Power Grid program (APG) identified and initiated in order to establish an Integrated ASEAN Power System between several countries comprising of islands: Malaysia, Thailand, Philippines, Cambodia, Myanmar, Vietnam and others			
	Leyte-Luzon in Philippines Interconnection Features: 440MW, 23km 350kV DC			
2000		*Completion of the Channel Islands Electricity Grid Project using XLPE cabling: France- Jersey (28km)–Guernsey (37km)*		

continued on following page

Table 2. Continued

Year	Asia	Europe	North America	Australia
2001	The Philippines enacted the Republic Act 9136 entitled as "Electric Power Industry Reform Act" in order to ensure and accelerate the total electrification of the country, security and affordability to the islanding region			
2003			The utility GRENLEC has initiated a comprehensive Interconnection policy. For distributed generation bind in the electrical grid for the island of Grenada in the Caribbean Islanding Region	
2004	The construction of a 2-staged interconnection project in Philippines was completed. Stage I, included the construction of a submarine cable between Leyte and Bohol by 69-kV lines. In Stage II, related transmission lines and substation facilities were constructed in 138 kV.		Agreed to establish a Task Force, Comprising several Caribbean Islands to develop recommendations for a Regional Energy Policy, which would address several issues of security, pricing & the impact on relative competitiveness in the CARICOM Single Market and Economy.	
2005	Cebu-Mactan Interconnection in Philippines (Features: 200 MW, 138 kV, 8.5 km) Enhancement of Leyte-Cebu in Philippines (Features: 230 kV, 17 km, 185 MW)			**The commissioning of the Basslink between Australia & Tasmania was put into operation (Features: 370monopolar, HVDC, 400 kV. The nominal rating of the link is 500 MW although it is capable of transmitting 630 MW**
2006	Panay-Boracay interconnection in Philippines (Features: 69kV, 1.6 km)			
2007	Enhancement of Cebu-Negros interconnection in Philippines (Features: 138 kV,18 km, 90MW)	*A 69 kV-cable with a capacity of 50 MW is commission in order to interconnection the power-grid of Formentera island with that of Ibiza. Also a 132 kV Line installed between Majorca & Minorca in the Balearic isles .*		

continued on following page

Table 2. Continued

Year	Asia	Europe	North America	Australia
2008	*A 33kV submarine cable to Koh Siboya, Koh Pu, Koh Phi Phi, Railey Bay, and Koh Lanta islands was submerged. (Features: total submarine 62.8 km, in two lengths, of XLPE insulated, lead sheathed, copper conductor 3-core 33 kV submarine cable, with an integrated fibre optic element)*	**The Interconnection project of Sardinia with Italy with two cables of total length 400km & capacity1000 MW is completed**	Hawaii published the Hawaii Clean Energy Agreement to transit Hawaii off of its imported fossil fuels reliance for electricity and transportation. The U.S. Department of Energy evaluated the interconnection plan of the electricity grids of the islands of Lanai, Molokai and Maui with Oahu.	
2009		DIRECTIVE 2009/72/ EC concerning common rules for the internal market in electricity & incorporating isolated systems forming electricity islands		
2010	ASEAN Interconnection Master Plan Study (AIMS) II agreed & published, prioritizing an number of future interconnections in the area of the South Eastern Asian Nations		World Bank establishes a strategy on Caribbean Regional Electricity Generation, Interconnection, and Fuels Supply.	
	The interconnection of Al Jasra (Bahrain), Ras Qurray (Saudi Arabia) sub stations, and Umman Na'san Island takes place (Features: 120km, HVAC, oil-filed, 400kV)		The United States Agency for International Development (USAID) publishes the Maldives Submarine Cable Interconnection Pre-feasibility Study, and proposed a number of interconnection projects in the Maldives islands complex.	
			Virgin Islands Water and Power Authority proposed a transmission project that interconnects the electrical power systems on the islands of: Puerto Rico (PREPA) St. Thomas and St. Croix (USVI) the British Virgin Islands (BVI)	

continued on following page

Table 2. Continued

Year	Asia	Europe	North America	Australia
2011	**ShangXiaChuan Island Interconnection with China (Features: 64/110kV, 30.293km)** The Taiwan Power Company commenced a project launched to manufacture and lay six submarine power cables between the island of Taiwan and the Penghu Islands, approximately 60 km off the southwest coast of the island of Taiwan.	**The Iberian Peninsula & Majorca Interconnection is commissioned (Features HVDC, 240km, 400MW, 250kV, Paper)**		
2012	**SiJiao island interconnection with the China mainland (Features: 64/110kV, 34.4kM)** **SiJiao island interconnection with the China mainland (Features: 64/110kV, 34.4kM)** **ZhouShan Island Interconnection with China is completed (Features: 2009-2012) project≈103 km.**	The European Economic and Social Committee publishes its Opinion "Getting EU energy islands connected: growth, competitiveness, solidarity and sustainability in the EU internal energy market" towards interconnections, sustainability & affordability in EU energy islands.	The World Bank assesses Regional Electricity Supply Options & Electricity market interconnections for the Caribbean Islands St. Kitts & Nevis (Caribbean Islands) interconnection with Puerto Rico study is launched in order to transfer large geothermal capacity located in the two islands	
2013			An amended CARICOM Energy Policy was approved alongside and the Caribbean Sustainable Energy Roadmap and Strategy for interconnections	
2014	The proposed interconnection of Borneo island (Malaysia) with Mindanao (Philipinnes) has been proposed for funding in the Asian Development Bank.	The Transmission System Operator of the Spanish Electricity System, Red Electrica de Espana (REE), awarded a contract for a double-circuit cable for the interconnection between the Balearic Islands of Mallorca & Ibiza (117 km, HVAC, 118 MVA)	**The Muskrat Falls Project is agreed including a hydro project of 824MW and the interconnection of Labrador (Canada) to Strait of Belle isle-Newfoundland island and Nova Scotia Island.**	

continued on following page

Table 2. Continued

Year	Asia	Europe	North America	Australia
2015	**Phu Quoc Island is interconnected to the national power grid in Vietnam (56 km of 3x630 mm² 110 kV copper, single wire armoured XLPE insulated submarine cable)** **The new submarine cable link from Guangdong power grid to Hainan island has been awarded (HVAC, 500 kV, single core, self-contained fluid filled cables, 32 600 MW capacity)**	*Tenerife and Gran Canaria and between Gran Canaria and Fuerteventura interconnection plan proposed in the EU*	The Ministry of Environment and Energy publishes the report for Greater Malé Region (Maldives) Renewable Energy Integration Plan with a detailed analysis of options for undersea electrical interconnections	
2016	**The Jindo–Jeju, submarine cable was commissioned to interconnect: South Korea (Features: 105 km HVDC, 400 MW, 250 kV)**	**St. George Island becomes interconnected with the Greek mainland (Lavrion) in order to transfer energy from a 73.2 MW wind farm built on the island (36.2 km)** *The Malta-Sicily Interconnection is commissioned (Features: HVAC, 220kV, 200MW, 98 km)*	**The Prince Edward Island's electricity interconnection completed (2*180MW) with Northumberland Strait and New Brunswick (Canada). The Project required to address the growing electricity demand & to deliver reliable, long-term energy for Islanders through energy imports.**	

Bold = Interconnection of one (or more islands) with the mainland
Underline =Interconnection of two islands
Regular =Interconnection of more than two islands between them
Italic =Policy, Regulation or Study/Report

APPENDIX 2: ISLANDS INTERCONNECTIONS

Table 3. Islands interconnections in Europe

Island(s)	Year	Country	Technology	Power (MW)	Voltage (kV)	Submarine Total Length	Cost (Mil €)	Other Technical Characteristics
Existing Projects								
Gotland I	1954	Sweden	HVDC	600	400	52	N/A	monopolar
SACOI (Sardinia-Corsica-Italy)	1968	Italy, France	HVDC	300	200	121	270	monopolar
Majorca-Minorca	1973	Spain	HVAC	80	132	42	N/A	4 *500 mm2, SCFF
Aland Islands	1973	Aland Islands, Sweden	HVAC	N/A	145	55	N/A	*Cross-linked polyethylene (XLPE)*
Gotland II	1983	Sweden	HVDC	130	150	92	N/A	monopolar
Normandie I (Jersey island)	1985	UK, France	HVAC	100	90	27	50	N/A
Gotland III	1987	Sweden	HVDC	130	150	92	N/A	monopolar
SARCO (Sardinia-Corsica)	1992	Italy, France	HVAC	150	100	16	38	bipolar
Normandie III (Jersey, Guernsey island)	2000	UK, France	HVAC	100	90	65	88,6	3-core cable, XLPE fiber optics
Formentera -Ibiza	2007	Spain	HVAC	50	69	22	N/A	N/A
Majorca-Minorca	2007	Spain	HVAC	-	132	40	N/A	N/A
RAMULO Project (The Iberian Peninsula &Balearic Islands)	2011	Spain	HVDC	400	250	237	420	bipolar
SAPEI (Sardinia)	2012	Italy	HVDC	100	500	420	730	bipolar
Aland Islands	2013	Aland islands, Finland	HVDC	200	80	158	130	HVDC light
Balearic Island of Mallorca & Ibiza	2015	Spain	HVAC	118	132	117	225	two three-phase cables with integrated fibre optics

continued on following page

Table 3. Continued

Island(s)	Year	Country	Technology	Power (MW)	Voltage (kV)	Submarine Total Length	Cost (Mil €)	Other Technical Characteristics
St. George Island -the Greek mainland (Lavrion)	2015	Greece	HVAC	75	150	36,2	36,4	N/A
The Malta-Sicily Interconnector	2016	Malta, Italy	HVAC	200	220	98	182	3x630mm2 submarine cables with XLPE insulation
Cycladic Islands Interconnection*	2016	Greece	HVAC	200, 140	150	234,6	400,17	3phase XLPE
Planned Projects								
Crete Interconnection (I)	2020	Greece	HVAC	400	150	120	328	N/A
Euro-Asia Interconnector	2022	Israel, Cyprus, Greece (Crete island)	HVDC	2000	N/A	1000	1500	N/A
Crete Interconnection (II)	2024	Greece	HVDC	700	N/A	340	1015	N/A

Data source: RTE, 2001; ABB, 2008; 4Coffshore; Nexant, 2010; Baltic Transport Journal, 2013; EuroAsia, 2013; Red Electrica de Espana, 2014; Ardelean & Minnebo, 2015; Govern Illes Ballears; Viohalco, 2015; Cirio, 2016; Paterno et al., 2016; Zafeiratou & Spataru, 2016.

Table 4. Islands Interconnections in Asia

Island (s)	Year	Countries	Technology	Power (MW)	Voltage (kV)	Submarine Total Length	Cost (Mil €)	Other Technical Characteristics
Existing Projects								
HVDC Hokkaidō–Honshū	1979	Japan	HVDC	300	250	44		bi-polar
Negros-Panay	1990	Philippines	HVAC	30	68	18	N/A	4 cables *18 Conductor 300, SCFF
Cebu-Negros	1993	Philippines	HVAC	90	138	18	N/A	
Java-Bali	1994	Indonesia	HVAC	N/A	150	N/A	N/A	
Leyte-Cebu	1995	Philippines	HVAC	185	230	32,5	N/A	Conductor 6 *30, SCFF
Haenam–Cheju between the Korean Peninsula and the island of Jeju in South Korea	1996	South Korea	HVDC	300	180	101	N/A	bi-polar
Penang Island	1996	Malaysia	HVAC	1000	275	14	N/A	6*14, Conductor 630, SCFF
Leyte-Luzon	1998	Philippines	HVDC	350	440	23	N/A	
Leyte - Bohol (Stage II)	2004	Philippines	HVAC	90	138	17	N/A	
Cebu-Mactan	2005	Philippines	HVAC	200	138	8,5	N/A	XLPE
Enhancement Leyte-Cebu	2005	Philippines	HVAC	185	230	17	N/A	
Panay-Boracay	2006	Philippines	HVAC	40	69	1,6	N/A	
Enhancement of Cebu-Negros interconnection	2007	Philippines	HVAC	90	138	18	N/A	
Koh Siboya, Koh Pu, Koh Phi Phi, Railey Bay, and Koh Lanta islands interconnection	2008	Thailand	HVAC	33	N/A	33,3	N/A	XLPE insulated, lead sheathed, copper conductor 3-core submarine cable, with an integrated fibre optic element)
The interconnection of Al Jasra, Ras Qurray and Umman Na'san Island	2010	Bahrain, Saudi Arabia	HVAC	1200 MVA	400	191	340	oil-filed
ZhouShan Island Interconnection with China is completed	2010	China	HVAC	N/A	64/111	6,9	N/A	1*500 mm, XLPE insulated, optic fiber
ZhouShan Island Interconnection with China is completed	2010	China	HVAC	N/A	64/112	13,8	N/A	1*500 mm, XLPE insulated, optic fiber

continued on following page

Table 4. Continued

Island (s)	Year	Countries	Technology	Power (MW)	Voltage (kV)	Submarine Total Length	Cost (Mil €)	Other Technical Characteristics
ShangXiaChuan Island Interconnection with China	2011	China	HVAC	N/A	64/110	30,3	N/A	N/A
Taiwan - Penghu Islands	2011	Taiwan	HVAC	N/A		60	N/A	N/A
SiJiao island interconnection with the Chinise mainland	2011	China	HVAC	N/A	64/110	34,4	N/A	1*500 mm, XLPE insulated, optic fiber
SiJiao island interconnection with the Chinise mainland	2011	China	HVAC	N/A	64/110	68,5	N/A	1*500 mm, XLPE insulated
ZhouShan Island Interconnection with China is completed	2012	China	HVAC	N/A	64/110	102,9	N/A	1*500 mm, XLPE insulated, optic fiber
Phu Quoc Island -national power grid in Vietnam	2015	Vietnam	HVAC	N/A	110	56	N/A	3x630 mm^2, copper, single wire armoured XLPE insulated submarine cable
Guangdong power grid to Hainan island	2015	China	HVAC	600	500	32	140	single core, self-contained fluid filled cables,
The Jindo–Jeju - South Korea	2016	South Korea	HVDC	400	250	105	N/A	N/A
Sumantra-Java	2017	Indonesia	HVAC	3000	500	35	N/A	N/A
Planned Projects								
Batangas-Mindoro	N/A	Philippines	HVAC	180	230	25	N/A	N/A
Mindoro-Semirara	N/A	Philippines	HVAC	180	230	34	N/A	N/A
Semirara-Panay	N/A	Philippines	HVDC	90	350	23	N/A	N/A

Data source: Indonesia East Java Electric Power Transmission and Distribution Network Project, 2002; Nexans, 2003; Nexant, 2010; Nexant, 2010; Orient Cable, 2011; Wu et al., 2013; ADB, 2013; Ardelean & Minnebo, 2015; TDP, 2015; TDP, 2015; Ardelean & Minnebo, 2015; TDP, 2015; Nai, 2015; Prysmian Group, 2015.

Table 5. Islands interconnections in North America

Island(s)	Year	Countries	Technology	Power (MW)	Voltage (kV)	Submarine Total Length	Cost (Mil €)	Other Technical Characteristics
Existing Projects								
Vancouver Island Pole 1	1968	Canada	HVDC	312	260	33	N/A	bi-polar
Prince Edward	1977	Canada	HVAC	138	200	30	N/A	3*240, SCFF
Vancouver Island Pole 2	1977	Canada	HVDC	370	280	33	N/A	bi-polar
Vancouver Island Pole 3	1984	Canada	HVAC	1200	525	39	N/A	4*1600 mm2, SCFF
Prince Edward	2017	Canada	HVAC	360	138	16,5	130	N/A
Planned Projects								
Labrador-Island Link	N/A	Canada	HVDC	900	315	35	N/A	N/A

Data source: Nexant, 2010; Ardelean & Minnebo, 2015; CBC, 2016; Stantec, 2017.

Table 6. Islands interconnections in Australia

Island(s)	Year	Countries	Technology	Power (MW)	Voltage (kV)	Submarine Total Length	Cost (Mil €)	Other Technical Characteristics
Existing Projects								
HVDC Inter-Island	1965	New Zealand	HVDC	1200	350	40	N/A	bi-polar
Basslink	2005	Australia	HVDC	500	400	290	1170	monopolar

Data source: Ardelena & Minnebo, 2015.

Chapter 3
A Global Perspective on Experiences and Practices for Low Carbon Technologies and Renewable Energy in Islands

ABSTRACT

Attention on islands energy systems is gradually increasing worldwide, enhancing sustainable resources on islands through a number of strategies and plans which aim to support and raise local awareness towards climate change. These plans, either in the form of official legal frameworks or through initiatives, aim at promoting energy efficiency, renewable energy, and effective resource management. Outcomes of those initiatives includes actions, programs, and projects where a number of islands demonstrate as test-beds for innovation and best practices. While autonomous states have larger control on implementing sustainable policies, they usually experience low economic prosperity. Islands that are part of a wider nation need to ensure adequate representation in government decision making. Due to the complexity of islands characteristics, best practices were discussed and analyzed for very small and small islands (micro), medium islands nations (meso), and large islands nations (macro).

DOI: 10.4018/978-1-5225-6002-9.ch003

INTRODUCTION

Island nations depending on their size, economic structure, population and location, encounter different barriers and drivers towards sustainable development. They require tailor-made policies and regulations to drive transition towards sustainable energy and resource management. These depends on various characteristics:

- *Climate* (cold, temperature, tropical) and *climatic parameters* (Meschede et al., 2016) (annual mean temperature, variance of monthly temperature, heating degree days, annual mean precipitation, coefficient of variation of monthly precipitation, annual mean global horizontal irradiance, variance of monthly global horizontal irradiance, annual mean wind speed) – islands located in colder climate have less tourists throughout the year
- *Socio economic parameters* (Meschede et al., 2016) – inhabitants, population, density, total griss domestic product, energy demand, economic activities
- *Physical characteristics* (Meschede et al., 2016; Sheldon, 2005) - area, the proximity to the mainland and other islands, the highest elevation of the island, the % of the island's area which is available for the implementation of large ground size renewable energy systems
- *Geographical restrictions* (Schallenberg-Rodríguez & Notario-del Pino, 2014) -– altitudes higher than 2000 m, hillsides with slopes exceeding 25%, protected areas, forests and woodlands with tree density greater than 25%, waters
- *The Governance of the island* (Sheldon, 2005) plays a key role in the support of sustainable development plans. This depends on if the island is an autonomous state or they are part of larger countries and follow the same national or regional policies could impact future plans based on the applied national strategies. While autonomous states have larger control on implementing sustainable policies, they usually experience low economic prosperity. Islands that are part of a wider nation need to ensure adequate representation in the government decision-making.
- *Population levels/Economic Growth* (Sheldon, 2005) is significant in designing sustainable policies. Islands have large population discrepancies thought the year. Usually, islands with low permanent population levels have weaker economies and cannot support large

volumes of tourists and fail to develop a robust tourism industry. Usually in such islands sustainable environmental practices in tourism can found prosperous ground even inside the local community while in larger islands centralized policies are required applicable to the different sub-sector of tourism

- *Homogeneity of the population and the socio-cultural sustainability of island goals* (Sheldon, 2005) affect the resilience of the locals to large tourism waves. Islands with homogeneous, indigenous populations are outstandingly helpless to tourism as they have established strong local values and practices and are resistant to changes. On the other hand, islands with more heterogeneous populations are more receptive to tourism and different cultures as well as sustainable development concepts and practices on that sector.

Each island offers unique opportunities for the implementation of renewable energies. Overviews on the current RES projects on islands can be found in Lynge Jensen, (2000); Neves et al. (2014). The latter reviews 28 research and demonstration projects on isolated and grid connected islands. It was concluded that many islands show convergences in their climatic, energy production possibilities and socio-economic structures. Despite this, a lack of information on the replicability and transferability of single island projects still remain obvious.

Initiatives aim at promoting energy efficiency, renewable energy and effective resource management. As an outcome of those initiatives includes actions, programmes, and projects where a number of islands demonstrate as test-beds for innovation and best practices. Table 1 provides a selection of islands nations who proved they could play the role as benchmarks and provide understanding of various practices under different conditions, which could be replicated in other cases. This selection was made following a review of over 520 islands using clustering analysis. The criteria used was:

- Islands should have already have played the role as test beds in research projects
- Different technologies and smart grid solutions have been tested and implemented
- Strong desire and support towards 100% renewable energy systems

Due to the complexity of islands characteristics, we analysed the best practices for very small and small islands, medium islands nations, large

Table 1. Categorisation of islands demonstrating various practices

		Isolated	Grid Connected Islands
Single Islands	Micro Islands Nations	Agios Efstratios El Hierro Utsira Ventotene	Bornholm Gotland Samsø Tilos Tokelau
	Meso Islands Nations	Barbados Crete Curaçao Grenada Guadeloupe Réunion Island	Jersey Mallorca
	Macro Islands Nations	Cyprus Jamaica	Malta Mallorca
Multiple Islands	Major Groups of Islands	Pacific Islands US Virgin Islands	
	Archipelagos	Faroe Islands Shetland Islands	

islands nations and I called them micro islands nations (Table 1) (referring to very small and small islands with a less than 5MW and 15 GWh energy required and population less than 100,000), macro islands nations (less than 35 MW and 100 GWh; population between 100,000 – 1mil.) and macro islands nations (> 35 MW and 100 GWh and population > 1mil.) and split them in isolated and grid connected islands. And also analyse best practices and challenges for major groups islands (e.g Pacific Islands, US Virgin islands) and archipelagos (e.g. Shetland islands, Faroe Islands)

The energy and electricity share for each of them based on the latest available data are illustrated in Figure 1 and Figure 2.

A detailed description of practices, research and demonstration (R&D) activities and innovation solutions towards 100% RES is provided in next section, as per each category highlighted in Table 1.

PRACTICES, R&D ACTIVITIES, AND INNOVATION SOLUTIONS WORLDWIDE: TOWARDS 100% RES ISLANDS

Islands are opportunity for demonstrating energy solutions. Islands' characteristics provide several advantages as laboratory for studying different degrees of renewable energies penetration into the grid. Some islands have

Figure 1. Selected islands and their energy share

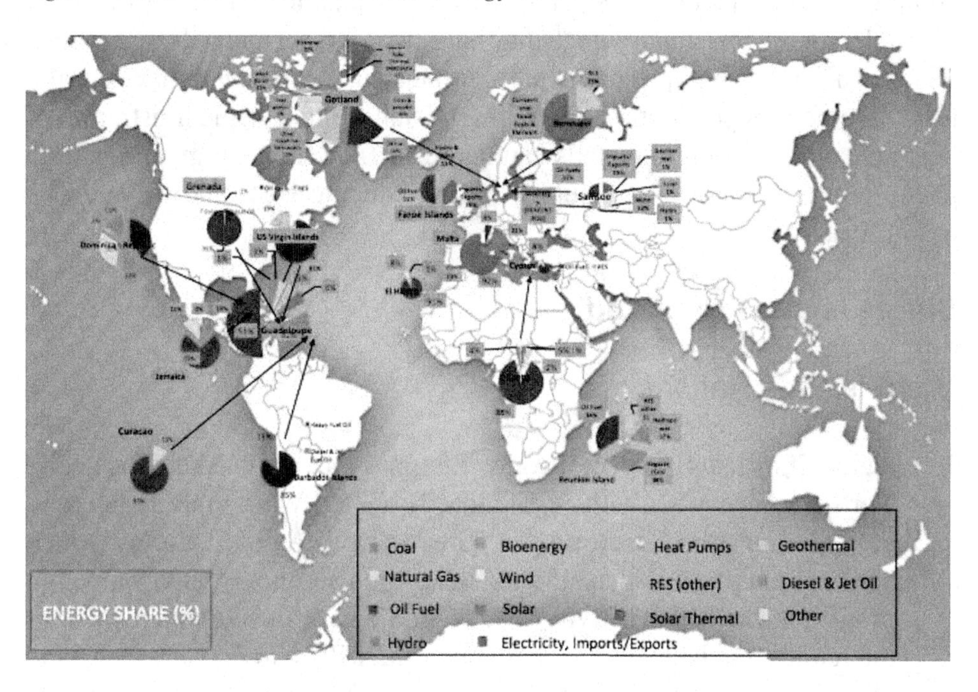

Figure 2. Selected islands and their electricity share

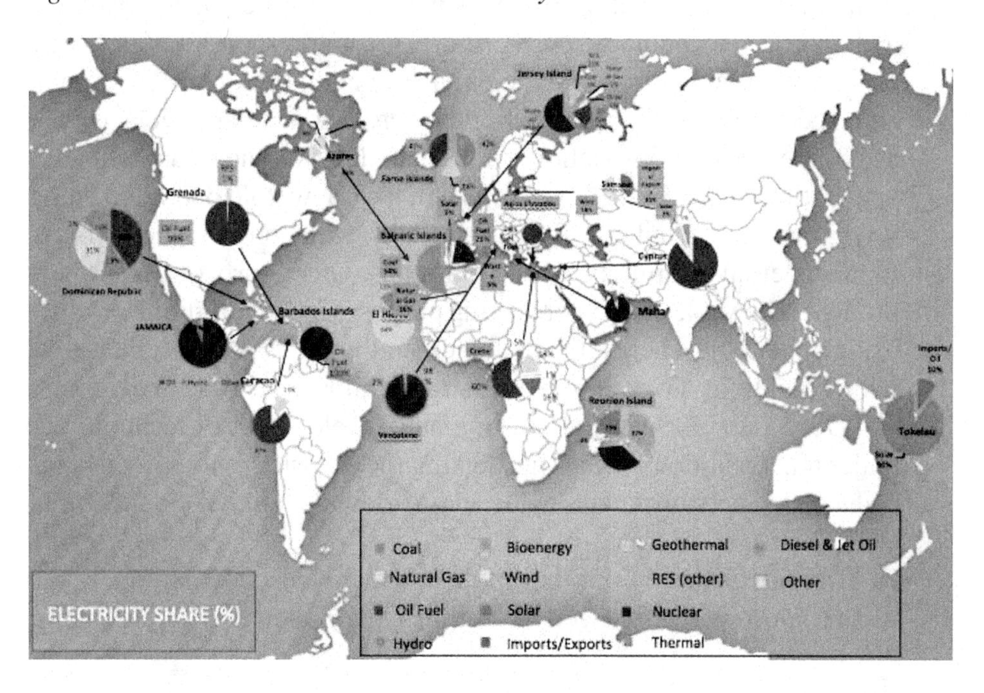

completely transitioned or are in the process of transitional phase. Transition is simpler for bigger islands which may have hydropower and volcanic islands which can develop pumped hydro and geothermal. Coral islands have usually less 'baseload' opportunities. The Canary Islands are a prime illustration of how an isolated group of islands can harness renewable energy and natural gas to reinforce their energy supply while reducing costs and dependence on oil fuels.

Single Islands: Isolated

Micro Islands Nations

1. **Agios Efstratios (Greece):** Agios Efstratios is a small island located in Northern Aegean Sea with a total capacity of 760 kW. Agios Efstratios will be used as a test-bed for the project "Green Island - Agios Efstratios" which aims to combine energy self-sufficiency and autonomy, environmental protection, research activity and technological development in non-interconnected islands. The main outcome will be to become the island 85% energy dependent from renewable energy sources and at a second step to become a 100% RES island. The project has a budget of 8.8 million euros, however still has not been implemented. The main components of this project are: wind turbines, solar photovoltaic panels, batteries and hydrogen technology. Furthermore, green transportation including electric vehicles, hydrogen vehicles and refueling stations. In particular, an electric vehicle charging station powered from solar panels as well as a hydrogen refueling station will be installed. Autonomous stationary applications in the form of fuel cells will have a twofold operation being used as uninterrupted power systems (UPS) while providing electricity to the external lighting of public buildings. Other parts of this project propose renewable energy for heating purposes in public buildings while using renewable energy technologies and energy saving schemes in public buildings. Overall, Agios Efstratios aims to become in the future an "Open Lab" for research, testing and development of technologies under real conditions while transferring the knowledge to other islands and isolated networks nationwide and worldwide (Nikitakos, 2009; Lin et al., 2016).
2. **El Hierro (Canary Islands, Spain):** The island of El Hierro is the westernmost and southernmost of the Canary Islands with an area of

268.71 km^2 (Instituto Geográfico Nacional, 2015) and a population of 10,892 (2003) inhabitants spread over small towns. The climate ranges from humid subtropical climate in the center to hot semi-arid and to a tropical desert climate in coastal parts.

El Hierro Cabildo (local government) had the idea of making the island an energy self sustained place. This possibility was supported in the "White Paper for a Community Strategy and Action Plan", developed by the European Commission (Gorona del Viento El Hierro, N/A). In 1997, the Council of the island of El Hierro approved El Hierro Sustainability Plan, aiming to make the island a self-sustained island. In 2000, the island was declared as a biosphere reserve, due to the actions of becoming self-supply of electrical energy via renewable energy. Under this framework, the Council of El Hierro, Unelco S.A. the Canary Islands Technological Institute have developed the project El Hierro Hydro-Wind Plant, aiming to transform the island in self-sufficient via renewable energy sources. In 2007, an agreement has been signed to guarantee the financing needed to implement the work. The project has been financed 60% with investments from public and 40% from private sector. Subsides from the Spanish General budget consist in the amount of EUR 35 million. The total cost of the project was 64.7 million euros. In addition a key part was the support from the public and this was done through a framework of the actuations carried out by the Government. These consist in energy saving, energy diversification, use of renewable energy, possibility of replication on other islands.

The wind-pumped hydropower station was put in operation since 2014. The project consists of a wind farm of 11.5 MW capacity as well as a hydroelectric power station of 11.32 MW capacity, also used for the production of desalinated water for irrigation activities (Kecskemeti, 2015). The wind farm supply power directly to the grid and simultaneously feed a pump unit that brings water into a higher reservoir tank, acting as an energy storage system. The hydroelectric plant uses the potential energy stored, ensuring energy supply and grid stability. The hydro-wind power plant of El Hierro is owned by Gorona del Viento, a public corporation owned by the Cabildo Insular del Hierro in 60%, the Canary Islands Government, through the Technological Institute of the Canary Islands in a 10%, and the utility Endesa, the energy supplier on the island. The power plant reached 46.5% of the total demand in 2017, reducing consumption of diesel by 8,000 tonnes (SteamGreen, 2018). In early

2018, El Hierro' electricity demand was covered entirely from renewable resources between 25th Jan and 12th February (SteamGreen, 2018).

El Hierro is one of the few islands that have already implemented a 100% Renewables Energy Plan. Additional projects include the production of biogas from waste under the initiative "El Hierro – zero waste" which comprises of the generation of biogas by utilizing stockbreeding effluents and sewage for methanogen aging as well as the RES establishment in the private sector and the presentation of an alternative transport system. On top, the authorities aim for improving the current plan, regarding the grid interconnection, in order to protect the local environment including the landscape and the ecosystem. This project wouldn't be successful enough without the use of energy storage through the hydro-pump power station of 6 MW pumping capacity with a system of artificial lakes (Eurelectric, 2012).

3. **Utsira (Norway):** Utsira is located in the North Sea, with 206 residents and having a 6- square-kilometre area. In 2004, the first full-scale hybrid project incorporating wind and hydrogen was developed on the island of Utsira in Norway by Statoil ASA and Enercon Gmbh. The main objective of this project was to demonstrate a safe hydrogen application which is able to deliver power in a secure and affordable way for the local houses in order to be able to commercialize this project in the near future (IPHE, 2011). Utsira presented excellent weather conditions for installing a wind energy system consisting of a wind turbine (600 kW), and a second wind turbine pitched-down to 150 KW connected to the hydrogen system comprising of a water electrolyzer (10 Nm3/h), with hydrogen gas storage (2400 Nm3, 200 bar), a hydrogen engine (55 kW), and a PEM fuel cell (10 kW) (Nakken & Ete, 2010). In order to provide reliability and minimize the impact of intermittency, the system has been equipped by a flywheel with a 5 kWh capacity and a 100 kVA master synchronous machine to balance and control voltage and frequency (IPHE, 2011). The system allows 2–3 days of complete energy autonomy (Nakken & Ete, 2010). Utsira is the world's first large-scale test of a stand-alone renewable energy system, using hydrogen as storage medium. The project has demonstrated that remote areas can be supplied with such systems and its lessons learnt could expand the market for future hydrogen. Furthermore, public awareness and acceptance could be improved through the demonstration of such projects.
4. **Ventotene (Italy):** Ventotene is a non-interconnected island using four diesel generators of 480kVA each. The island suffers from monthly

demand fluctuations due to increased tourism levels during the summer in parallel with decentralized energy growth mainly in terms of rooftop solar installations.

A project demonstrating best practice was installed in 2015 comprising of 300kW/600kWh lithium-ion batteries, which allow for higher levels of solar energy penetration into the island. The project aims to improve the flexibility of the grid while minimizing the operation of the local oil-fired engines and improving the island's sustainable plan. The energy storage system, partially funded by EIB has succeeded to reduce the island's fuel consumption requirements by 15% and respectively 15% in the annual CO_2 emissions reduction. In parallel, it has allowed to increase distributed renewable hosting capacity.

In 2016, on Ventotene, Enel unveiled a system that integrates generation and storage to provide an island's energy supply. It is a hybrid operation, depending on a 600 kWh lithium-ion battery to bring an improvement in the quality of service offered to residents. The island is a tourist destination with high peaks of demand for electricity during the summer season. Diesel generators have to be turned on and off frequently, with a non-optimal load that causes an increase in fuel consumption and affects the overall efficiency of the system. The storage system helps to store any excess electricity produced at peak times, and managed to supply 39% of islands load when is needed. The project demonstrates an innovative mechanism in the sense the battery system is coupled to diesel generators as it embeds an advanced control system incorporating both battery and diesel parallel operation. This system enhances efficiency as during short-term peak loads power is supplied by the battery system and not from the generators while in periods of low demand there could be an option to switch off the diesel generators entirely. Through this mechanism the operator manages to reduce fuel consumption and prevent maintenance (Eurelectric, 2017). Such approach can be replicated to many other islands worldwide.

Meso Islands Nations

1. **Barbados:** Barbados is an island country in the Caribbean region of North America. Barbados has taken the lead in the US and globally in the installation of solar water heaters, ranking fourth worldwide in installed capacity per capita in 2012 (NREL Barbados, 2015).

Since the early years of 1970s Barbados established a series of measures to promote alternative resources and efficiency as a consequence of the oil crisis. One outstanding early demonstration was a solar water heater installation on the Prime Minister's residence, which led to a 70% reduction in annual gas consumption. Since then, hundreds of projects have been implementing in different kind of buildings and sectors with multiple benefits and currently Barbados managed to reach penetration of solar water heaters of approximately 40%. In total, based on an approximate estimation the total installed capacity of solar water heaters on the island has reached 88 MW, more than half of the island's historical peak electricity demand. Although Barbados islands have a significant potential for exploiting RES resources, growth in the near future so far no large scale renewable energy project has been implemented while the total installed capacity of PV systems reached 3.5 MW in 2014 approximately half of the total allowed capacity on the island (NREL Barbados, 2015). In addition, Barbados has recived 24.6 Million dollar funding part of the Public Sector Smart Energy Programme (PSPP), funded by the Inter-American Development Bank (IDB) and the European Union (EU). The project aims to replace all street lights with Light Emitting Diode (LED) fixtures. In addition intends to expand the generation of electricity from renewable energy sources with the aim to reach 100 percent reliance on renewable energy by 2045 as per BL&P 100/100 Vision.

2. **Crete (Greece):** The island of Crete is the largest non-interconnected island in Greece, located in the southern part of the Aegean Sea and consumes more than 50% of the total power consumed in the Aegean Sea non-interconnected islanding region (Hellenic Electricity Distribution Network Operator, 2014). Electricity demand is covered by three local thermal power plants with a total capacity of 823.46 MW. The predominant fuel is heavy fuel oil (HFO), and secondly diesel fuel, which is used to cover peak demand as it records considerably higher prices and increased tax rates. The steam turbines and the CCGT station are mainly used to cover the base load, due to their slow response to demand fluctuations while gas and diesel turbines provide power during the peak hours. Due to the increased use of oil, the carbon emissions intensity factor for Crete is 0.69 t/MWh, which is 12% higher than the average national carbon intensity factor and produces approximately 1.8-1.9 million carbon tones from electricity production per year. Renewable energy sources (RES) represent 24% of the total annual electricity

generation, which is the largest share in the non-interconnected Greek region at this moment. The renewable energy capacity consists of wind and solar energy projects (200.3 MW and 78.3 MW respectively) and a small hydropower station (0.3 MW) (Hellenic Electricity Distribution Network Operator, 2016) . Other forms of energy such as bioenergy, solar thermal and hybrid technologies have not been exploited yet for the electricity sector, however there are installations providing energy for heating purposes with an approximate generation of 360 GWh per year. Also, 455,000 m^2 of solar thermal collectors provides about 2% of the total final energy consumption (Lin et al., 2016). The island has significant potential to expand its renewable energy installations while implementing energy efficiency projects due to its high level of consumption in the commercial sector (including mainly tourism). Also due to high wind curtailments the island requires new investments in energy storage systems.

3. **Curaçao:** Curaçao island is located in the southern Caribbean Sea, about 65 km (40 mi) north of the Venezuelan coast. Curaçao published in 2009 an energy policy document, which puts forward a general framework and principles to improve the energy conditions on the island. In addition, it sets principles to reduce energy consumption by 40% by 2020 and supports the demonstration of combining wind power with storage to supply the total island's energy needs (NREL Curacao, 2015).
   This policy was updated in 2011 to support the introduction of more independent power producers to the system while reducing generation costs. Furthermore, it provided incentives to limit import taxes on renewable energy generation equipment while creating a tax credit for the installation of those systems. Also in 2011 a net metering program for distributed wind and solar generation systems was set up. In 2015 the program was replaced with a feed-in tariff considering all building installations (NREL Curacao, 2015).
   Curacao, has, an excellent wind potential and already installed 58.5 MW of wind power plants. However, it does not feature any utility-scale solar installations up to now but has as of now being introduced a few distributed solar installations, with the total interconnected capacity reaching 8 MW (ECOENERGY, 2014).
4. **Grenada:** The island of Grenada is located northwest of Trinidad and Tobago, northeast of Venezuela and southwest of Saint Vincent and the Grenadines. The island has set a number of environmental and clean energy goals for supporting renewable energy sources while promoting other

policies such as demand-side energy efficiency schemes, programmes to supplant incandescent lights with fluorescent lights including also an energy efficiency program, which targets a 10% reduction in government electricity use. The island of Grenada was part of the Global Environment Facility's two-year Energy for Sustainable Development in Caribbean Buildings Project. In addition, Grenada demonstrates a few small scale renewable energy and energy efficiency projects such as (NREL Granada 2015).:

a. a 80-kW wind turbine built in 2007
b. the Maca Bana Villas 111 panel and 10-kW solar PV system built in 2009
c. the demonstration of 1 kW of wind and 1.8 kW of solar PV, 148.5 kW of PV power at Grand Anse developed in 2013
d. the 31.6-kW ground-mounted PV system
e. and few other roof-mounted PV systems installed around the island

Furthermore, The Grenada Electricity Services Limited has further supported the development of RES projects with a strategy to spend $150 million USD towards this direction. Some of the future projects include two geothermal plants which have the potential to meet almost the total power requirements of Grenada (NREL Granada 2015). Further developments could be made. For example, the Electricity Supply Act has not been updated to address self-generation and distributed renewable energy.

5. **Guadeloupe:** The energy mix in Guadeloupe is dominant by oil fuel. The island of Guadeloupe has published two key papers for regulating its energy policy: the Regional Plan for Renewable Energy and the Rational Use of Energy (PRERURE) and the Regional Climate, Air and Energy Scheme (SRCAE). PRERURE, established in 2007, was updated in 2012 including as a priority the creating of a platform for policy recommendations from different energy sectors.

   In addition, PRERURE formed the island's RES target for 50% share in the electricity sector by 2020 and by 2030 the target aims to reach 50% of the whole primary energy sector. In contrast, the SRCAE is a broader paper which explores air pollution and greenhouse gas emissions reduction; while it is aligned with the PRERURE policy. Another policy raised as an outcome of the two aforementioned policies is the Thermal Regulation of Guadeloupe, which proposes a number of energy efficiency measures for buildings such as energy consumption standards for new buildings, obligatory energy efficiency certifications that have to be renewed every 3 to 10 years, and energy feasibility studies.

Guadeloupe has already installed several renewable energy projects with a total capacity of 112.8 MW including wind, solar and hydro projects. An innovative project deployed by EDF is the EDF Energie Nouvelles, which has installed several solar and wind projects, without being the only investor as several other companies have deployed energy storage systems on the island (BP Solar, Tenesol, and Alinea Solar in PV24; AEROWATT and SEC in wind). Also, a geothermal project has been installed by Geothermie Boulliant. In terms of bioenergy applications, Albioma operates a 56.5 MW power plant, which uses bagasse as a feedstock after the sugarcane harvest but as an alternative could also utilize coal

Due to the fact that Guadeloupe as the rest of the non-interconnected islands has as an upper limit the 30% max RES integration the local authorities investigate solutions to raise this threshold. A project developed by EDF named as "Millener," begun in 2012, is an innovative project that will provide to local consumers who have PV systems appliance-connected energy controls or home energy storage systems to maximize their efficiency while reducing costs and enabling demand response and increasing grid stabilization (NREL Puerto Rico 2015).

6. **Réunion Island:** The island of La Réunion is located in the Indian Ocean and belongs to France. Energy consumption is mainly dependent on conventional fuels such as oil and coal. However, the local communities and local authorities aim for a energy transition towards cleaner energy forms. This was further supported by a French regulation called "Energy transition for a green growth" (Eurelectric, 2012) which fully supports fuel dependency for the island by 2030. Also the "PEGASE" project was established on the island to support renewable energy. Through this project, asodium/sulphur (NaS) battery of 1 MW/7 MWh was installed on the island to improve the intermittency originated from renewables while allowing higher levels of RES penetration to the system and while balancing the system through day-ahead and intra-day forecasting of generation for the two main PV farms (OFARnR, 2016).

   A similar project is currently under implementation by the Direction des Systèmes Energétiques Insulaires (SEI) (EDF, 2016) which is developing two energy storage systems in the French Antilles, based on an alternative battery technology (less sophisticated than NaS), with a power capacity of 5 MW.

The Reunion island, has moreover made an intercommunity which has as a primary objective to plan, fund, create and work a cold generation and district cooling network utilizing deep seawater and "SWAC" (sea water air conditioning) innovation. That innovative network will supply with air conditioning local buildings. The total capacity of the project is 40 MW including offshore and distribution networks as well as a pumping station, that will utilize seawater pumped up from great depths (Eurelectric, 2017).
The total cost including both the design and the implementation of the project is estimated at 135€ million. While using the SWAC system, the local residents will manage to achieve energy efficiency reaching up to 80% (Eurelectric, 2017).

Macro Islands Nations

1. **Cyprus:** Cyprus' population is approximately 750,000 having an isolated power system without interconnection. It depends almost entirely on imported fuels. The average selling price (prior taxation) of diesel and petrol is systematically higher than the comparing EU27 prices. Since 1st January 2009 the electricity market has been liberalised for all non-domestic customers. Legislative Framework for RES was ordered in 2003: The revenues (almost 22eurosM) of this finance are coming from the consumers paying an extra tax of 0,44 eurocents/kWh. More about support schemes and grid issues can be found on the Legal sources on renewable energy website. Cyprus operates three local power generation plants with a capacity of 1477.5 MW, while renewable sources are limited to 8% of the total installed capacity (Zachariadis & Hadjikyriakou, 2016). A number of programmes were established from the Cypriot Energy Regulatory Authority (CERA) and the Ministry of Energy (MoEICT), Electricity Authority of Cyprus (EAC) and the Distribution System Operator (DSO) which allowed for several micro-scale projects to become interconnected in the national grid. In particular, the "Green+, Zero Energy Mountains of Cyprus" programme has as a main goal to manage and optimize renewables distributed generation in rural areas of PVs installed on low voltage network (20 MW) and medium voltage network (50 MW). Also, the project "SmartPV8, smart net metering for promotion and cost-efficient grid-integration of PV technology in Cyprus" has as a key objective to improve and support the Net-metering scheme and consequently increase the RES share in the electricity mix.

Another initiative, the e-charge is an electric vehicle charging service launched by the EAC for electric vehicles (EVs), in order to be able to access the EAC owned charging infrastructure in public areas. This was a pilot-project consisting of 16 charging stations installed at various public spots and in order to support and promote e-mobility on the island. Project oriented, the Electricity Authority of Cyprus licensed a 50 MW solar thermal power plant of parabolic trough type, anticipated to become operational in the near future. The project has a thermal energy storage capacity of 850 MWh using heat transfer fluid. In best conditions the thermal energy storage will be completely charged within 6 hours during the summer months. The total investment cost for the solar thermal power plant is estimated at €360 million expected to become financed by the European Investment Bank.

However, Cyprus suffers from a chronic shortage of water. It relies heavily on rain to provide household water, but according to Department of Meteorology (2015) in the past 30 years average yearly precipitation has decreased. Since 2001, water desalination plant has been constructed to supply almost 50% of domestic water. The plan for Cyprus is to increase the renewable electricity share from wind power by 2020. In addition, to be part of the EuroAsia Interconnector project, which aims to connect Israel, Cyprus and Greece with 2000 MW HVDC undersea power cable (European Commission, 2017).

2. **Jamaica:** Jamaica is an island country situated in the Caribbean Sea, being a leader in the transition to sustainable energy systems in the Caribbean. In 2009 the National Energy Policy was published (Ministry of Energy and Mining, 2010), part of which the main targets were set for the island of Jamaica in terms of renewable energy share, energy efficiency and emissions reduction for the milestone year of 2030. In that sense, Jamaica, aims to reduce carbon intensity significantly by 50% from 2015 to 2030. Energy efficiency measures proposed by the local government propose tax exemptions for energy efficiency equipment, energy labeling for refrigerators and freezers, and utility-led energy audit programs. Additionally, the government aims to decrease by 30% the energy costs for public buildings. On this basis, in the private building sector, the National Building Codes were restructured in 2009 to include the International Building Codes, which include the updated requirements for energy consumption and conservation.

Although Jamaica is doing great efforts towards renewables, its current matrix of renewable power plants comprises approximately 7% of total installed generation capacity.
Moreover, the Jamaica Public Service Company Limited set a Net Billing program which has allowed for bill credits at the utility's avoided cost rate for the energy surplus exported to the grid by customer-sited generation. Up to now, 80 systems have been installed, with a total capacity of 1.4 MW, while more than 250 local consumers have applied for participation in the local Office of Utilities Regulation for the Net Billing program (Jamaica Observer, 2015).
Jamaica now hostesses 41.7 MW of wind energy farms split into two projects and 9 hydroelectric power plants with a total 30 MW capacity. However, solar development is still limited on the island despite the potential with only 32 distributed generation small and medium scale projects.
In November 2012, a capacity of 115 MW of renewable energy projects became available from the local authorities following the submission of proposals to acquire part of the capacity. In 2014, the local authorities confirmed the contracts with three counterparties for the energy-only portion of the request for proposals. The winning projects were: BMR Wind Jamaica, which will construct 34 MW of wind at a cost of $90 million, Wigton Wind Farm, which will extend its existing farm by 24 MW a for $46 million and WRB Enterprises, which will execute a 20-MW solar facility for $60 million (NREL Jamaica, 2015)

Single Islands: Grid Connected Islands

Micro Islands Nations

1. **Bornholm:** Bornholm is a small sized island in the middle of the Baltic Sea, active mainly in agricultural activities, while demonstrating a range of best practices. Bornholm has shifted already away from fossil fuels and relies solely from energy by wind, sun and biomass. It is predicted to be carbon neutral by 2025. To do this, the local DSO did setup 1,000 new photovoltaic systems and fuels the local plants only with locally produced biomass. Households' energy sourcing was converted from oil-fired burners to district heating, heat pumps and solar heating.

Bornholm also invested in charging stations for electric mobility and launched a campaign to promote car-free transportation.

The first community initiative came in 2007 under the name 'Bright Green Island'. A year later, the Municipality of Bornholm established the target to transform the island to a carbon neutral community by 2025. In 2015, the strategic energy plan towards this target became more concrete through a collaboration of the local authorities and utilities. The pioneering simulation of the local energy system incorporating data analysis for all the sectors is an innovative, unique project demonstrating a best practice on the island. Currently, two projects are being implemented on the island: the Strategic Energy Plan which is based on the aforementioned plan, to transform Bornholm to a CO_2-neutral community based on sustainable and renewable energy by 2025 (Aegean Energy Agency, 2016).

The second project is the EcoGrid 2.0 which is the extension of the EcoGrid EU, completed in September 2015 and received the prestigious EU Sustainable Energy Award (EUSEW16) in 2016. This project consists of 15 partners and the main objective is to expand the range of products provided to the energy customers while in parallel enhancing system's reliability and flexibility as well as renewable energy capacity and mainly wind in parallel with energy storage and smart grid equipment in the local distribution network. This could be achieved through a real-time price signal which managed to reduce the peak load by approximately 670 kW, also emphasis was placed on the heating sector with equipment which can control their heating system (EcoGrid, 2017). The project will cost 48.6 million DKK and will be financed from the Danish Energy Technological Development and Demonstration Program, co-funded by the European Union (Aegean Energy Agency, 2016). One of the challenge in the transition process is the management of the volatility of renewables' power plants and the flexible consumption of certain technologies (e.g. heat pumps).

2. **Gotland Island:** Gotland is the largest Swedish island. The island is covered in proportion of 40% with forests and 31% of land area used for grazing and arable land[4] The total transited power in the local grid during 2016 was 919.061 (979.75 GWh including losses). Power generation costs have the same national price since Gotland is interconnected to the grid. The main users of fossil fuels are the cement industry and transportation.

New technologies have been installed in Gotland Island. This includes industrial energy recovery in the local cement factory in Slite which consumes approximately 30% of the total power generation transmitted to the island the rest is equally divided between the city of Visby and the rest of Gotland. The reserve capacities come from local diesel generators to serve the system in instances of power outage in the mainland grid connection. In terms of transmission infrastructure, the commissioning of the world's first commercial HVDC transmission link was implemented to connect Gotland and the Sweden mainland (Features: 20 MW, 150 kV, 96 km). Furthermore, the enhancement of the Gotland-Sweden Cable to 130 MW was the first in the world to feature a fully redundant digital control and protection system as well as a gas insulated switchgear (GIS). The first cable of 1954 was replaced in the Gotland-Sweden Interconnection. A bipolar link has been formed, while the total operational transmission capacity is 260 MW, but it has a max capacity of 320 MW.

3. **Samsø (Denmark):** Samsoe located in the Central Denmark exemplifies through several renewable energy projects of wind, solar and bioenergy technologies. Although the island still relies in a small extent on fossil fuels, mainly for transportation or in periods of low wind speeds, Samsø has achieved an annual average of 100% RES supply (Lin et al., 2016). Until 1997, Samsø depended entirely on oil and coal imports. In 2007, an offshore wind farm of 10 turbines funded by islanders was completed. Homes are heated with straw burned in a central heating system and they power some vehicles on biofuel. The surplus of electricity from wind energy is exported to the mainland grid.
   The initiative to transform the island to a green energy hub initiated in 1997 when the island was awarded to become a 100% RES island. Despite the fact that Samsø has already attained the net clean exports from the island to the mainland to become higher compared to the imports from mainland, recently a more ambitious new plan was adopted the "Samsø 2.0" where the local residents will become actively involved in the island transition (Aegean Energy Agency, 2016). The core of this project is Samsø to become fossil fuel free in all the sectors (electricity, transport, industry, agriculture) by maintaining and upgrading the old facilities while placing emphasis on implementing new projects. Furthermore, energy conservation in heating is proposed with the target to reduce energy for heating purposes by 30% by 2020. This will be achieved by a district

heating supply system using local biomass resources and through the development of an individual heating system using sustainable energy forms. For industry, the energy consumption for heating will be reduced by 5% in 2020. Also, savings in the electricity sector have been included in the plan. All these targets will be accomplished through partnerships and joint ventures with the strong support of the local community (Saastamoinen, 2009; Lin et al., 2016). The island has now moved to the "Samsø 3.0" phase which deals with the circular bio economy, that could have a significant impact in small and medium-sized enterprises on Samsoe island (Energy Academy, 2017).

4. **Tilos Island:** Tilos is a small Greek island located in the Aegean Sea, being part of the Dodecanese group of islands. The island has an extraordinary biodiversity with underground springs that feed five wetlands. Tilos Project initiated in 2015 demonstrates one of the key innovative sustainable projects currently under implementation European wide. Tilos recently received the first prize at the EU Sustainable Energy Week (EUSEW) contest in both the Energy Islands and the Citizens' Awards categories. The main objective of this project is the complete energy dependence of the island. Currently, the island is interconnected and power is being transmitted from the power station located on the island of Kos (Chatzivasileiou, 2017). The project consists of an 800 kW wind turbine and photovoltaic panels with an output of 160kW. Additionally, a trial product storage system for storing the surplus energy on very sunny or windy days until it is needed in battery with the following characteristics: NaNiCl2 battery storage of 2.88MWh/800kW and advanced battery inverters. The project incorporates a smart meter and demand side management (DSM) equipment to provide the required energy monitoring and short-term energy management in local residents and centralized loads such as pumping stations. Finally, an Energy Management System (EMS) that will coordinate operation of the whole TILOS system has been proposed (Chatzivasileiou, 2017). TILOS island will manage to achieve through a micro-scale grid energy management project, approximately 100% RES penetration while enhancing the network stability and flexibility. The surplus of the produced energy will be exported to the neighboring islands while the interconnection with Kos will facilitate as an option to provide ancillary services to the interconnected network with the neighboring islands (Tilos Horizon Project, 2017).

5. **Tokelau:** Several islands belonging to the Pacific nations are shifting towards renewable energy to meet their total energy requirements, due to the increased impact of the climate change at these regions (Newlands, 2015). For instance, Tokelau a small Polynesian island country, managed to increase significantly its renewable energy capacity demonstrating a best practice at a global level. Before 2012 Tokelau islands relied on three diesel-driven power stations.
   Despite the low number of inhabitants managed to achieve high levels of wind energy penetration, with solar energy being the primary energy source. Currently the island has three solar photovoltaic system one on each atoll. The 4,032 solar panels with one MW capacity, 392 inverters, 1,344 batteries provide 150 percent of their current electricity demand. As back up they use generators which run on local coconut oil.
   The initial tender specification called for the solar systems to supply 90% of Tokelau's electricity demand, based on announcements from the New Zealand-based project contractor, PowerSmart, the island managed to exceed this already ambitious target with a total cost of £6m (NZD $12.5 million) (Newlands, 2015; Wilson, 2017). The main objective for the implementation of this plan is to use the savings from fuel costs to support other sectors on this island such as education and health.

Meso Islands Nations

1. **Jersey Island:** Jersey island demonstrates a number of innovative projects. Beyond the submarine interconnections with France (in 1985 and 2000) (4Coffshore, 2018; Electricity Grid System Manager RTE, 2001 that allowed to decommission significant capacity of the local thermal oil-fired power plants, in 2008 the government of Jersey initiated the implementation of a new energy from waste (EFW) power plant.
   This is considered so far, one of the biggest projects ever implemented with high capital intensity equal to approximately 130€ million. This installation is important as it offers a secure, naturally friendly framework of processing the municipal waste and combustible solid waste produced on the island. The processing waste produces steam as a by-product and this steam is coordinated to a conventional steam turbine and electrical generation unit with a 10MW generator in order to produce power. The modern plant utilizes a two stream moving grate incinerator joining a heat recovery and gas cleaning system. GHG emissions are limited with

the use of a dry gas scrubbing mechanism with activated carbon and lime injection and urea-based NO_x abatement.

Another initiative by the local government is related with boosting electric vehicles use. This will be implemented by installing chargers in public parking spaces or while providing further incentives to the local drivers and vehicle owners in collaboration with the local car dealer network which should be committed to bring electric vehicles onto the island (Eurelectric, 2012).

2. **Mallorca:** Mallorca is the largest island in the Balearic Islands. Electric vehicles have found fertile ground to develop in the Balearic island of Mallorca through the "ecaR project". Building on a previous project which developed the first electric charging network for electric vehicles including six fast strategically installed charging "ecar" spots, it is the first Fast Self-Charging Club developed. Ecar, is a European funded project which provides an electric vehicle network for the island. Users will be able to make use of a smartphone application which enables drivers to identify the closest charging point as well as other relevant information (Eurelectric, 2017).

    Beyond electric mobility, Mallorca has limited local oil-fired generation (Figure 4) as the interconnectors with the neighboring islands of Menorca and Ibiza and the Spanish peninsula through the RAMULO project completed in 2011, supply the main power loads to the island (Kasselouri et al., 2011; CPMR,2017).

Macro Islands Nations

1. **Malta:** Malta is a southern European island country. Malta initiated the development of an innovative close-to-shore offshore wind project with a capacity of 95 MW in 2009. The project will contribute in accomplishing the national renewable energy share of 10% in gross final energy consumption for Malta according to Directive 2009/28/EC (European Union, 2009). The Initial capital investment cost for this project is evaluated to be between €280 and €335 million (Eurelectric, 2012).

    Furthermore, in order to phase the challenge related to deep waters, another offshore project with a total capacity of 54 MW will take place with the use of a pioneering floating wind farm. This innovative platform will have the capability of rotating while taking advantage of the optimum angle for wind speeds and directions, in order to maximize generation. It is estimated to cost €200-250 million (Eurelectric, 2012). The European Union Directive 2009/28/EC set Malta's target share of renewable energy at 10% by the year 2020. The mandatory 10% target for transport concern also Malta.

One of the most significant projects for Malta's energy system is the Malta-Sicily interconnector with a length of 95-kilometre, 230 kV and a total cost of 182 million euros (Paterno et al., 2016), which expanded Maltas' limited energy market access to the Italian and consequently European energy market while enhancing interconnectivity among EU Member states. As a result, the local thermal power operation has been significantly limited by 324 ktoe (European Commission, 2016) and a new modular and highly efficient combined cycle diesel engine plant capable of burning both heavy fuel oil and gasoil, has been replaced the older ones. This new power plant is equipped with modern flue gas abatement technology, being able to abate emissions of nitrogen oxides (NO_x), sulphur dioxide (SO_2) and particulate matter (PM), following Directives 2010/75/EU and 2015/2193/EU (European Union, 2010; European Union, 2015). In addition, using the waste heat recovery boilers, fuel consumption per kWh generated is reduced and extra capacity is attained, with a plant efficiency raise of 46.7% . Finally, Malta hostesses a large number of small scale PV installations, currently having a nominal capacity of approximately 80 MW and an effective annual penetration of 3.3% (Eurelectric, 2017). There are also a few waste-to-energy generators.

2. **Mallorca (Spain):** Electric vehicles have found fertile ground to develop in the Balearic island of Mallorca through the "ecaR project". Building on a previous project which developed the first electric charging network for electric vehicles including six fast strategically installed charging "ecar" spots, it is the first Fast Self-Charging Club developed. Ecar, is a European funded project which provides an electric vehicle network for the island. Users will be able to make use of a smartphone application which enables drivers to identify the closest charging point as well as other relevant information (Eurelectric, 2017).

   Beyond electric mobility, Mallorca has limited local oil-fired generation as the interconnectors with the neighboring islands of Menorca and Ibiza and the Spanish peninsula through the RAMULO project completed in 2011, supply the main power loads to the island (Kasselouri et al., 2011; CPMR,2017).

Multiple Islands

Major Groups of Islands

1. **Pacific Islands:** The three major groups of islands in the Pacific Ocean are Polynesia, Micronesia and Melanesia. Melanesia island are large, mountainous and mainly volcanic islands. Polynesian and Micronesian islands are mid-sized island states with limited resources, little or not commercial forests and mineral deposits, but enjoy a high standard of living from foreign assistance and remittances from expatriate island communities.
   Beyond Tokelau, other islands in the Pacific region have expressed their wish to become 100% RES in the near future. Islands such as Fiji, Papua New Guinea, the Solomon Islands and Vanuatu possess a large potential for solar, wind, hydro and geothermal energy. Also, Samoa has set a concrete target to achieve 100% renewable energy by 2017. Projects will be funded by a consortium that will consist of the states of China and New Zealand as well as the Asia Development Bank.
   The Cook Islands have also set the target to become energy independent by 2050. While recently, a $20.5 million project was implemented on the Cook Islands installing a number of solar PV projects around the islands (Wilson, 2017).
   Furthermore, a $100 million contribution to the Samoa-based Green Climate Fund, to be utilized for a variety of mitigation and adaptation projects counting renewable energy development. In the setting of the energy transition initiative in the Pacific islands states the Prime Minister of Samoa announced a new Pacific Climate Change Centre (PCCC) funded by Japan. The PCCC will build up collaborations all through the pacific islands states and globally to distinguish transition procedures for Pacific countries to climate change (Wilson, 2017).
2. **US Virgin Islands:** The U.S. Virgin Islands' Clean Energy Goals are to reduce fossil fuel consumption by 60% in 2025 while enabling a 30% max output from renewables (EIA, 2016) . These targets are aligned with the Act 7075 passed in 2009. According to this Act, renewable energy development will boost through a number of different measures such as: net metering schemes for distributed generation and obligatory installation of energy-efficient solar water heating systems for every new building by the year of 2020. Also, the Virgin Islands Energy

Office managed to accumulate $32 million from the Recovery Act to fund discounts, grants and loans for energy efficiency and distributed renewable energy technologies. Furthermore, a regulation framework was established in 2014, which configured the Feed-In Tariff Act for renewable energy projects participating in the market.
In terms of already implemented projects, one of the biggest PV system (comprising of a 448-kW photovoltaic (PV) system) was installed at the Cyril E. King Airport on St. Thomas in the Caribbean region. Furthermore, several energy efficiency refurbishments took place in 11 school buildings leading to energy cost-savings of $1.3 million for the first year and $1.7 million for the second year, which led the local government to create a $35 million fund in 2013 to install lighting and water retrofits in 34 more schools. Finally, several other projects have already been implemented since February 2010 including a 1,500 solar water heating and PV systems throughout the territory and a 15 MW of distributed photovoltaic panels have already been installed or they are under implementation. Potential projects for the future, are biomass projects on the island of St. Croix with a capacity of 3-5 MW as well as landfill gas. In addition to woody biomass utilization, a 7-MW anaerobic digester using king grass as a feedstock has signed a Power Purchase Agreement (PPA) with the local operator. In particular, the U.S. Virgin Islands have a potential to cover from 8 MW to 33 MW from waste treatment (NREL US Virgin Islands 2015). The government goal is to reduce fossil fuel consumption by 60% from 2008 levels in all consuming sectors by 2025 (Lantz et al., 2011). Finding sites for wind turbines has been proven difficult, as well as the risk of damage from hurricanes. This has made financing wind projects difficult (Virgin Islands Energy Office, 2009). There is also potential for waste-to-energy, landfill gas and biomass energy. However, waste-to-energy plants at landfills on St. Thomas and St. Croix has been proven to be too costly (Nowakowski, 2016).

Archipelagos

1. **Faroe Islands:** The Faroe Islands are located at an archipelago located in the north Atlantic sea, between Scotland and Iceland, consisting of 18 small islands, 17 of which are populated. Faroe islands have the status of an autonomous, self-governing country within the Kingdom of Denmark

(Eurelectric,2012). The islands are windy, cloudy and cool throughout the year. Sunny days are rare. Energy is produced from fossil fuels, hydro and wind power. In 2016 a 2.3 MW 700 kWh lithium-ion battery became operational (Thomson, 2016). Furthermore there are plans to convert the existing hydropower to pumped-storage hydroelectricity, because rain and wind are high in winter and low in summer (SEV, 2017). Tidal power is also considered.

Faroe islands are developing under ambitious targets of a zero carbon electricity sector by 2030 while proposing rapid electrification of the heating and transport sectors. These islands have achieved high levels of renewable energy integration equal to 42% for hydro, 18% for wind and the remaining 40% from fossil fuel. As such, Faroe islands have to confront with a number of difficulties in the operation of the local electrical system e.g. short circuit, inertia, stability etc. Consequently, the Faroe Islands, exemplify as test-beds for new technologies in order to manage the increased levels of stochastic resources.

Specifically, the EU-funded TWENTIES PowerHub project has already been applied in the islands successfully. Through this project industrial consumers participate in a load-shedding scheme, decoupling loads like heat pumps, cold storages and freezing compressors when a local frequency deviation is occurred (Eurelectric, 2012).

In 2016, a 2.3 MW/0.7 MWh battery system was installed in order to enhance stability for a 11.7 MW wind farm output. The outcomes of the battery use proved that it has achieved to manage fluctuations of the generation output. Further projects are on progress in order to support decarbonization on the Faroe islands. For example, a Quick-Charger infrastructure project is currently explored with several charging points across the islands that will support electric vehicles growth. The main objective is to increase in parallel renewable energy resources that will charge EVs through the excess of their energy output in parallel with the installation of energy storage systems (Eurelectric, 2017).

2. **Shetland Islands:** The Shetland Islands comprise of 16 individual islands with a total area of 1,468 km^2 and a total population of 23,210 (National Records of Scotland, 2012). The Shetland Islands comprise more than a hundred islands, but only fifteen of them inhabited. The Shetland Isles receive at least 2 mm of rainfall for 250 days of the year (Citizens Advice Bureau, 2013). The Islands are not connected to the UK mainland electricity network. Electricity supply comes from three

main energy sources. Islands residents can choose their suppliers exactly as UK residents on the mainland.

Given the islands isolated position and cold climate, the islands population consumes more than double the average consumption in Scotland. Shetland residents consume 10,384 kWh in comparison to 4,863 kWh, the average household sale in Scotland (Shetland Islands Council, 2011). The higher energy demand from customers is largely due to extreme climate and cold temperature meaning that heating is required for longer periods throughout the year. This high demand has led to fuel poverty existing on the islands. Fuel poverty is defined as when a customer has to be spend greater than 10% of their income on fuel bills. The figure is thought to be as high as 40% on the Shetland Islands (Citizens Advice Bureau, 2013).

The PURE Project installed in the Shetland Islands, is an independent, small scale hybrid system which exploits wind energy in parallel with hydrogen with a total cost estimated at 400,000£ (Smallest, N.P.Programme and E.Union, n.d.). This is a project founded by public and private parties supported also by the EU. This project aims to demonstrate the efficient use of wind in parallel with hydrogen in order to cover energy demand in distant and isolated areas. The main objective of the PURE Project (Smallest, N.P.Programme and E.Union, n.d.) was to demonstrate the safe and effective use of hydrogen. Hydrogen is produced from electrolysis powered by a Wind Turbine which saves energy and transforms it to power through fuel cells. This system includes a control centre for local and distant operation. This system comprises of a W/T of 500 kW capacity and electrolyses of 25 kW operating at 25 bar. Also, hydrogen storage of 40 N m3 in MH tanks, a 1-stage H2 compressor at 220 bar, and one filling station at 220-bar bottles complete the system's equipment (Smallest, N.P.Programme and E.Union, n.d.). One of the conclusions drawn from the project was that the conventional alkaline electrolyser technology to produce hydrogen is not very effective when using an intermittent energy supply such as wind.

FUTURE RESEARCH DIRECTIONS

Currently, many islands worldwide pay extremely high energy prices, due to import of fuel at long distance. Renewable energy can provide multiple benefits, on one hand lower decrease fuel expenditures, on the other hand lower carbon emissions. There are already islands which demonstrated turn towards renewables. Such examples include Floreana in Galapagos, El Hierro in the Canary Islands, Gotland in Sweden, Tuvalu, Cape Verde, Cook Islands, Hawaii and so on. Through all these practices islands prove that generating electricity from an increasingly diverse array of sources make their systems more reliable.

However, more research and demonstration is needed to analyse the stability of the grid in terms of frequency and voltage control, reactive power supply when high renewable energies are integrated in the system. The impact on climatic conditions should also be examined and test industrial equipment under different conditions. Under normal conditions, the equipment could work well, but when implemented outdoor in a different environment could be proved to be difficult and expensive to maintain it.

Examining the power demand pattern in a more detailed way (hourly) with advanced metering systems could help understand patterns.

Thus, it is clear that there are challenges and further examinations to be understood and solve, but in the same time there are many opportunities for islands towards more efficient and effective operation of power systems in islands, which should be analysed.

Creating an inventory of existing information and recent research and compiling a comprehensive guide on islands worldwide will help understand the diverse characteristics of islands. In addition, key performance indicators that could be used to assess the level of the energy transition as well as to monitor the progress should cover different aspects:

- Energy and environment (e.g. consumption per inhabitant, sources of production, share of renewables, seasonality, level of interconnection, greenhouse gas emissions, energy poverty, etc.)
- Regulatory (e.g. existence of specific frameworks for the island which prevent / favor the energy transition such as national laws, regulations or restrictions, and Operational Programmes from ESIF)
- Financial (e.g. surplus cost of electricity compared to the mainland and potential use of the differential for investments, existence of incentives and funding specific for the island, subsidies, etc.)

The existing of an integrated sustainable energy action plan could also be helpful. The success of such a task consist in accessibility and existing of data. The study should be supplemented by a deepened analysis of a statistically significant number of islands that allows for extrapolation.

CONCLUSION

Islands energy systems at micro, meso and macro level, either interconnected or not have been analysed. A number of problems have been identified. Disproportionately high and inelastic operator guarantee costs, keep independent suppliers away from non-interconnected islands markets. Another problem arise due to the extensive use of fossil fuels, the high concentration of emissions pollutants. Moreover, islands have very particular characteristics due to their reliance on tourism as their principal source of income. As such, the evident discrepancies in energy demand between summer and winter could cause power failures. Some islands have to conquer additional challenges such as the shortage of fresh water leading to the use of desalination plants, with higher power costs. On top of that, problems related to waste management and transport complete a series of structural problems contributing to the continuing degradation of those regions.

Taking into consideration the above, worldwide efforts have been made targeting low carbon energy solutions, by increase use of renewables and replace of large, old fossil fuels plants with low and zero carbon technologies. Several programmes have been launched moving towards a more renewable predominant economy on islands. Various islands projects are in progress or have already been implemented. A catalyst towards the promotion of renewable energy technologies agains all odds could be a united effort from the government, municipalities, local communities, businesses, NGOs and academia to provide systematic support and guidance towards sustainability and technological development in order to improve local economies, tourism, quality of life.

The outcome of this chapter is to compile current information and provide a basic understanding platform for addressing future concerns in today's decisions and assist in meeting long-term challenges in generating new knowledge. One key conclusion from the analysis is the adoption of renewable energies and interconnection with mainland or neighboring islands could benefit an island by improving energy security and lowering its carbon footprint. It may also allow for the export of surplus energy to mainland and neighbouring islands.

REFERENCES

4Coffshore. (2018). *Normandie 1 (N1) Interconnector*. Retrieved from http://www.4coffshore.com/windfarms/interconnector-normandie-1-(n1)-icid78.html

Aegean Energy Agency. (2016). *Smart Islands Projects and Strategies*. Friedrich Ebert Stiftung. Retrieved from http://library.fes.de/pdf-files/bueros/athen/12860.pdf

Carpio, C., Coviello, M. F., Horta, L. A., Peña, J., Gamarra, A., & Santana, B. (2010). *Energy efficiency in Latin America and the Caribbean: Situation and outlook*. Retrieved from https://www.cepal.org/publicaciones/xml/2/39412/lcw280i.pdf

Chatzivasileiou, A. (2017). *Double distinction for Greek island of Tilos in EU energy awards*. WWF. Retrieved from http://www.wwf.gr/en/news/1979-double-distinction-for-greek-island-of-tilos-in-eu-energy-awards

Comision National De Energia. (2012). *Status and Outlook of the Renewable Energies and Energy Efficiency in the Dominican Republic*. Global workshop on clean energy development: Establishing a Foundation for Low Carbon Energy Systems, Washington, DC. Retrieved from https://www.usea.org/sites/default/files/event-/Dominican%20Republic%20Country%20Presentation_0.pdf

Council of the European Union, European Parliament. (2015). *Directive (EU) 2015/2193 of the European Parliament and of the council of 25 November 2015 on the limitation of emissions of certain pollutants into the air from medium combustion plants*. Retrieved from https://publications.europa.eu/en/publication-detail/-/publication/dab34984-9560-11e5-983e-01aa75ed71a1/language-en

CPMR. (2017). *Personal Communication with the Conference of the Peripheral Maritime Regions* (CPMR). Retrieved from http://cpmr.org

ECOENERGY. (2014). *Aqualectra bevindt zich op een kruispunt*. Retrieved from: http://www.ecoenergycuracao.net/aqualectra-bevindt-zich-op-een-kruispunt/

EcoGrid. (n.d.). *Information and education of the future electricity consumers: Experience from EcoGrid EU on Bornholm*. Retrieved from: http://www.eu-ecogrid.net/

EDF. (2016). *EDF dans les territoires insulaires*. Retrieved from: https://www.edf.fr/groupe-edf/premier-electricien-mondial/strategie/edf-dans-les-territoires-insulaires

EIA. (2016). *US Virgin Islands Territory Profile and Energy Estimates*. Retrieved from: https://www.eia.gov/state/?sid=VQ

Electricity Grid System Manager (RTE). (2001). *Presentation of the second Jersey-France Interconnection*. Retrieved from http://clients.rte-france.com/htm/an/journalistes/telecharge/dossiers/jersey_france_an.pdf

Energy Academy. (2017). *Samso 3.0*. Retrieved from: https://energiakademiet.dk/en/2-0/

Eurelectric. (2012). *EU Islands : Towards a Sustainable Energy Future*. Retrieved from https://www3.eurelectric.org/media/38999/eu_islands_-_towards_a_sustainable_energy_future_-_eurelectric_report_final-2012-190-0001-01-e.pdf

Eurelectric. (2017). *Towards the Energy Transition on Europe 's Islands*. Retrieved from http://www.elecpor.pt/pdf/20_02_2017_Eurelectric_report_towards_the_energy_transition_on_europes_islands.pdf

European Commission. (2017). *EuroAsia Interconnector – Final Detailed Studies Prior to Project Implementation, Part of Cluster of the Projects of Common Interest 3.10*. Retrieved from https://ec.europa.eu/inea/sites/inea/files/3.10.1-0004-cyel-s-m-16_action_fiche.pdf

European Commission Directorrate – General for Energy. Directorate-General for Climate Action and Directorate-General for Mobility and Transport. (2016). *EU Reference Scenario 2016 Energy, transport and GHG emissions Trends to 2050*. Retrieved from https://ec.europa.eu/energy/sites/ener/files/documents/20160713%20draft_publication_REF2016_v13.pdf

European Union. (2009). *Directive 2009/28/EC of the European Parliament and of the Council of 23 April 2009 on the promotion of the use of energy from renewable sources and amending and subsequently repealing Directives 2001/77/EC and 2003/30/EC*. Retrieved from http://eur-lex.europa.eu/legal-content/EN/TXT/PDF/?uri=CELEX:32009L0028&from=EN

European Union. (2010). *Directive 2010/75/EU of the European Parliament and of the Council of 24 November 2010 on industrial emissions (integrated pollution prevention and control)*. Brussels: Official Journal of the European Union. Retrieved from http://www.prtr-es.es/data/images/Nueva-DEI-EN.pdf

Gorona del Viento El Hierro. (n.d.). *The Wind-Hydro-Pumped Station of El Hierro*. Retrieved from http://www.goronadelviento.es/index.php?accion=articulo&IdArticulo=121&IdSeccion=104)

Grassi, S., Chokani, N., & Abhari, R. S. (n.d.). Large scale technical and economical assessment of wind energy potential with a GIS tool: Case study Iowa. Energy Policy, 45, 73–85.

Hellenic Electricity Distribution Network Operator. (2014). *Monthly Reports of RES & Thermal Units in the non-Interconnected Islands-September*. Author.

Hellenic Electricity Distribution Network Operator. (2016). *Monthly Reports of RES & Thermal Units in the non-Interconnected Islands-January*. Author.

IEA. (2016). *Energy Policies of IEA Countries: Portugal 2016 Review*. IEA.

Instituto Geografico Nacional. (2015). *Islas e islotes con superficie superior a 1km2*. Author.

IPHE Renewable Hydrogen Report. (2011). *Utsira Wind Power and Hydrogen Plant Utsira Island*. Retrieved from http://www.newenergysystems.no/files/H2_Utsira.pdf

Jamaica Observer. (2015). *Gov't reports strong take-up of net billing licences*. Retrieved from http://www.jamaicaobserver.com/news/Gov-t-reports-strong-take-up-of-net-billing-licences_18351721

Kasselouri, B., Kambezidis, H., Konidari, P., & Zevgolis, D. (2011). Environmental, economic and social aspects of the electrification on the non-interconnected islands of the Aegean Sea. *Energy Procedia*, *6*, 477–486. doi:10.1016/j.egypro.2011.05.055

Kecskeméti, G. N. (2015). *Renewable energy integration in Small and Isolated Power Systems in Spain (SIPSS). Case study of the hydro-wind power station on El Hierro Island*. Conference on Sustainable Energy for SIDS by ESMAP, Vienna, Austria. Retrieved from http://esmap.org/sites/esmap.org/files/DocumentLibrary/4b%20-%20Gabriella_SIPSS%26ElHierro_17June2015_Optimized.pdf

Krulewitz, A., & Litvak, N. (2013). *Solar in Latin America & The Caribbean 2013: Markets, Outlook & Competitive Positioning*. Retrieved from: https://www.greentechmedia.com/research/report/solar-in-latin-america-the-caribbean-2013

Lantz, Olis, & Warren. (2011, September). *U.S. Virgin Islands Energy Roadmap: Analysis, National Renewable Energy Laboratory*. NREL/TP-7A20-52360 Executive Summary.

Lin, J., Wu, Y., & Lin, H. (2016). Successful experience of renewable energy development in several offshore islands. *Energy Procedia*, *100*, 8–13. doi:10.1016/j.egypro.2016.10.137

Lynge Jensen, T. (2000). *Renewable energy on small islands* (2nd ed.). Copenhagen: Forum for Energy & Development.

Mathiesen, B. V., Hansen, K., Ridjan, I., Lund, H., & Nielsen, S. (2015). *Samsø Energy Vision 2030: Converting Samsø to 100% Renewable Energy*. Copenhagen: Department of Development and Planning, Aalborg University.

Meschede, H., Holzapfel, P., Kadelbach, F., & Hesselbach, J. (2016). Classification of global island regarding the opportunity of using RES. *Applied Energy*, *175*, 251–258. doi:10.1016/j.apenergy.2016.05.018

Ministry of Energy and Mining. (2010). *National Renewable Energy Policy 2009-2030 ... Creating a Sustainable Future*. Retrieved from http://mset.gov.jm/sites/default/files/pdf/Draft%20Renewable%20Energy%20Policy.pdf

MSET. (2015). *An Overview of Jamaica's Electricity Sector*. Retrieved from: http://mset.gov.jm/overview-jamaicas-electricity-sector

National Renewable Energy Laboratory (NREL). (2015a). *Energy Transition Initiative, Energy Snapshot, Puerto Rico*. Retrieved from https://www.nrel.gov/docs/fy15osti/62708.pdf

National Renewable Energy Laboratory (NREL). (2015b). *Energy Transition Initiative-Energy Snapshot Grenada*. Retrieved from https://www.nrel.gov/docs/fy15osti/62699.pdf

National Renewable Energy Laboratory (NREL). (2015c). *Energy Transition Initiative-Energy Snapshot U.S. Virgin Islands*. Retrieved from https://www.nrel.gov/docs/fy15osti/62701.pdf

National Renewable Energy Laboratory (NREL). (2015d). *Energy Transition Initiative-Energy Snapshot Jamaica*. Retrieved from https://www.nrel.gov/docs/fy15osti/63945.pdf

National Renewable Energy Laboratory (NREL). (2015e). *Energy Transition Initiative-Energy Snapshot Barbados*. Retrieved from https://www.nrel.gov/docs/fy15osti/64118.pdf

National Renewable Energy Laboratory (NREL). (2015f). *Energy Transition Initiative-Energy Snapshot Curacao*. Retrieved from https://www.nrel.gov/docs/fy15osti/64120.pdf

National Renewable Energy Laboratory (NREL). (2015g). *Energy Transition Initiative-Energy Snapshot -Dominican Republic*. Retrieved from https://www.nrel.gov/docs/fy15osti/64125.pdf

Neves, D., Silva, C. A., & Connors, S. (2014). Design and implementation of hybrid renewable energy systems on micro-communities: A review on case studies. *Renewable & Sustainable Energy Reviews*, *31*, 935–946. doi:10.1016/j.rser.2013.12.047

Newlands, M. (2015). Pacific Islands heading for 100% renewable energy. *The Ecologist*. Retrieved from: http://www.theecologist.org/News/news_round_up/2887157/pacific_islands_heading_for_100_renewable_energy.html

Nikitakos, N. (2009). *Green Island Agios Efstratios*. Retrieved from http://www.infostrag.gr/syros/wp-content/uploads/2009/06/NIKITAKOS_SEMINARIA-ERMOUPOLHS-20091.pdf

Nowakowski, K. (2012, May 22). Growing Pains: Large-Scale Composting in the Virgin Islands. *St. Thomas Source*.

Oeystein, U., Nakken, T., & Arnaud, E. (2010). The wind / hydrogen demonstration system at Utsira in Norway: Evaluation of system performance using operational data and updated hydrogen energy system modeling tools. *International Journal of Hydrogen*, *35*(5), 1841–1852. doi:10.1016/j.ijhydene.2009.10.077

Office franco-allemand pour les énergies renouvelables (OFAEnR). (2016). Smoothing an intermittent generation: Interest of generation forecast and storage global management. *Prévision des Énergies Renouvelables et Garantie Active Par le Stockage d'Énergie*.

Paternò, G., Madonia, A., Ippolito, G., Massaro, F., Favuzza, S., & Energia, D. (2016). *Analysis of the new submarine interconnection system between Italy and Malta : simulation of transmission network operation*. 2016 IEEE 16th International Conference on Environment and Electrical Engineering (EEEIC), Florence, Italy. 10.1109/EEEIC.2016.7555467

Red Electrica de Espana. (2016). *Balearic Islands' electricity system*. Retrieved from: http://www.ree.es/en/activities/balearic-islands-electricity-system

Red Electrica de Espana. (n.d.). *Data*. retrieved from https://demanda.ree.es/movil/canarias/el_hierro/total

RSE. (2014). *Report on small Italian islands not interconnected with the mainland*. RSE.

Saastamoinen, M. (2009). *Case Study 18: Samso - renewable energy island programme*. Changing Behaviour, project co-funded by the European Commission within the 7th Framework Programme Theme Energy 2007.9.1.2. Energy behavioural changes. Retrieved from www.energychange.info./downloads/doc_download/337-cbcase18-denmarksamsoe

Schallenberg-Rodríguez, J., & Notario-del Pino, J. (2012). Evaluation of on-shore wind techno-economical potential in regions and islands. *Applied Energy*, *124*, 117–129.

SEV. (2017). *Tangible plan for the green course*. Retrieved from http://www.sev.fo/Default.aspx?ID=193&Action=1&NewsId=2921&PID=392

Sheldon, P. J. (2005). The Challenges to Sustainability in Island Tourism. *Occasional Paper, 2005*(October), 1. Retrieved from: http://citeseerx.ist.psu.edu/viewdoc/download?doi=10.1.1.502.5607&rep=rep1&type=pdf

Smallest N. P. Programme and European Union. (n.d.). *Promoting Unst Renewable Energy (PURE) Project*. Retrieved from: http://www.smallestnpp.eu/documents/PUREprojectcasestudyforinternet.pdf

Smilegov Multilevel Governance. (2013). *SMILEGOV Enhancing effective implementation of sustainable energy action plans in European islands through reinforcement of smart multilevel governance*. Islands Strategy Deliverable D 6.1. Submitted by CPMR, D6.1 Islands Strategy Paper. Retrieved from: http://www.greenpartnerships.eu/wp/wp-content/uploads/Islands-strategy-communication-paper.pdf

States of Guernsey. (2011). *Guernsey Energy Resource Plan*. Retrieved from https://www.gov.gg/CHttpHandler.ashx?id=5575&p=0

Thomsen, B. (2015). *Renewable energy developments in the Faroe Islands*. Presentation at Island Energy – Status and Perspectives, Tokyo, Japan. Retrieved from: https://www.iea.org/media/workshops/2015/egrdoct/12Thomsen_Jarofeingi.pdf

Thomson, C. (2016). *Lithium-ion batteries can help to safeguard the grid*. Retrieved from https://eandt.theiet.org/content/articles/2016/11/lithium-ion-batteries-can-help-to-safeguard-the-grid/

Tilos. (n.d.). *Optimum Integration of Battery Energy Storage, Horizon 2020-Low Carbon Energy – Local/ small-scale storage LCE-08-2014*. Retrieved from https://www.tiloshorizon.eu/images/deliverables/TILOS-Flyer_EN.pdf

Virgin Islands Energy Office, Office of the Governor. (2009, May 5). *U.S. Virgin Islands Comprehensive Energy Strategy*. Author.

Wilson, C. (2012). Pacific Island Sets Renewable Energy Record. *Sustainable Energy for all*. Retrieved from http://www.ipsnews.net/2012/10/pacific-island-sets-renewable-energy-record-2/

Zachariadis, T., & Hadjikyriakou, C. (2016). State of the Art of Power Generation in Cyprus. In *Social Costs and Benefits of Renewable Electricity Generation in Cyprus, Springer Briefs in Energy* (pp. 7–16). Cham: Springer. doi:10.1007/978-3-319-31535-5_2

Chapter 4
Business Models and Policies to Support Renewable Energy Community in Islands

ABSTRACT

This chapter starts with an introductory part, explaining the role of business models and analyzing the different financial models of ownership. It has been concluded that in renewable energy projects, ownership business models center in specialized complexity, economies of scale, capital costs, and financing perspectives based on its own characteristics. It has many favorable features including the ability to provide power to local communities and create jobs. However, business models include several decisive financing, service, and monitoring characteristics. Business models should be dynamic, while being adjusted to the special conditions, features, and risks of the given project. In renewable energy projects, ownership business models center on the specialized complexity, economies of scale, capital costs, and financing perspectives. The public-private partnership (PPP) is usually the optimum business model option for medium- to large-scale or grid-connected renewable energy projects, and is usually applied with a structure of a built-own-operate-transfer or multiparty ownership.

DOI: 10.4018/978-1-5225-6002-9.ch004

BUSINESS MODELS AND RENEWABLE ENERGY PROJECTS

In general, business models (BM) define how investments will be planned, designed, implemented, and managed. Business models include several decisive financing, service, and monitoring characteristics. The scope and the size of each project have to be well defined, as well as the target group. Business models should be dynamic, while being adjusted to the special conditions, features and risks of the given project. The business model that fits the best to a given project varies upon a number of parameters such as the local conditions, the financial and regulatory framework, the institutional context and the support mechanisms that may exist. Business models are usually structured upon the basis of the service considerations the scale of the project, the consumer, and the regulations.

Business models can be sketchily classified into two categories as presented below (ADB, 2015):

- *Ownership models*, which emphasize on financing and risk mitigation concerns
- *Service models* which center on giving indicated services and highlight distinctive strategies of operation and support

In reality, actual business models are a combination of several forms and methods.

In renewable energy projects, ownership business models center on the specialized complexity, economies of scale, capital costs, and financing prospects of each project based on its own characteristics. The public–private partnership (PPP) is usually the optimum BM option for medium- to large-scale or grid-connected renewable energy projects, and is usually applied with a structure of a build–own–operate–transfer (BOOT) or multiparty ownership. It has been shown that in practice, small-scale projects include lease or hire purchase and dealer credit sale models (Roehrich et al., 2014).

The service-based business model is considered to offer either a product or a service to the consumer. Usually, the benefit is conveyed by an energy service company which can work as private or public utility, a cooperative, a nongovernment organization (NGO), or a private entity. These entities are part into two fundamental groups: an energy supply contracting company and an energy performance contracting company.

The most common business model in this category is the fee for service. Under this business model, the user pays a fee based on usage or energy savings. The most common case is that beneath the standard utility service contract, energy consumers will pay a tariff for electricity as specified from the national market operator. Be that as it may, in the case of isolated systems, where the infrastructure and generation resources must to begin with be set up (e.g. a microgrid, interconnections, renewables), a user cooperative business model may be more effectively chosen over other business models (ADB, 2015).

In general, renewable energy projects face a number of challenges as every business venture. Renewable energy projects such as wind, solar and biomass although considered as mature technologies still encounter difficulties associated with risks stemming from the aforementioned characteristics as well as specific details of the project as defined by the geographical and techno-economic specifications of the project. Business models are called to address technical and technologies challenges, administrative, market and financial challenges during RES projects implementation (Pasicko, 2016b).

Renewables have multiple benefits for isolated power systems as islands, since they reduce subsidized, expensive fuel imports while they support economic activities, they improve the quality of life, they reduce emissions, they support sustainable tourism and they assist in preserving the local environment. A key element however towards successful renewable energy deployment on islands is community involvement. Community involvement can take different forms as it can facilitate either projects for self-consumption or investments in projects which fed power into the local grid. Through the involvement of the local community and the development of the ownership feeling more renewable energy projects can be integrated into the system, while local expertise is developed for supporting O&M, although challenges have to be addressed in terms of access to knowhow and expertise (Roberts *et al.*, 2007). The ownership might range from 100% for small sized projects to lower rates, through co-owner-ship arrangements where third parties' private companies or the public sector participate as the majority shareholders (Leaney *et al.*, 2001). However, to support community ownership public awareness and education are required. It is noticed, according to Rogers et al. (2012), based on an interview sample that no interviewer declared willingness to become a project leader while participating actively in the coordination of a renewable energy project at a community level.

Projects that do involve community ownership have different legal and financial models of ownership as described in the following (Walker, 2008):

- **Cooperatives:** Where local citizens in the neighbourhood community become members of the cooperative and purchase shares to fund the project. Or in other words, organizations that incorporates citizens in decision-making processes (Rijpens et al., 2013), delivering services through democratic membership (Viardot, 2013).
- **Crowdfunding:** Is comparative as the cooperatives structure but the investors might come from anyplace in the world and is usually facilitated through an online platform (Pasicko, 2016a)
- **Community Charities:** That usually act as associations of NGOs with charitable status that gives or runs facilities for the local community.
- **Development Trusts:** Have been primarily utilized to speak to communities' interface in revenue-generation enterprises, and in certain circumstances this has been extended to integrated alternates of community ownership.
- **Shares Owned by a Local Community:** Where companies give shares of a project for free to a local community organisation or to the local municipalities. This practise has been utilized as a way of giving a community advantage that is closely tied to the execution of the generation unit (Centre For Sustainable Energy, 2009)
- **Part-Ownership:** By the community which may give restricted rights to control or to make inputs into decision making

These projects and their financial models of ownership are discussed in the next section.

The development of renewable energy projects and their business models can vary across regions, due to institutional rules, physical conditions, energy policies that provide grid access, political decision, cultural and social norms (Bauwens et al., 2016). Furthermore, synergies between sectors must be created. Electric vehicles are unlikely to reach their full potential in islands unless new business models rise that produce new connections between private drivers, islands authorities, energy suppliers will be created. Using EVs as a power source or power sink can offer a much broader business model as the passive storage function. The e-mobility business models have to work over the boundaries of these systems, and bring innovation through use of new commodities and services across these three large systems (Bolton & Hannon, 2016; Hall & Roelich 2016).

FINANCIAL MODELS AND OWNERSHIP FOR RENEWABLE ENERGY COMMUNITY PROJECTS

There are different types of participation and possibilities to invest in renewable energy projects. These include: Bond, Cooperative, Donation, Generation kWh, Loan and Reward.

A loan and bond are similar in the sense both allow you to lend money to a project, but the difference consist in the way the repayment takes place. In the case of a loan, the payback is done in periods with interests rate over a period of time, while in the case of a bond, the project repay your investment in either one single transition at the end or over time.

Cooperatives offer investment opportunities either through equity or other investment types. Energy cooperatives usually are formed of members in the local community willing to invest their money to support projects such as solar rooftops, wind turbines, biogas plants and later to benefit from the energy they produced (Citizenenergy, 2016). Cooperatives require the purchase of a share or "non-withdrawable deposit" to their membership. This entitles the members to vote and to participate democratically in the cooperative (Citizenenergy, 2016). When a member leaves the cooperative, the stock/deposit is redeemed at the original purchase price by the cooperative. They usually have a local, regional or national scope and imply a long-term investment. However, members can leave the cooperative but still get the deposit /common stock back but the redemption must be taken place within a time defined. In this case, the risk of investment is low and it usually depends on cooperative's credit quality, revenue model of RES projects developed (such as feed-in-tariff), the type of energy produced and so on.

Crowdfunding is another concept for sustainable energy projects. It is actually an extension of the cooperative idea but it applies to larger communities. Anyone in the world can invest money via internet. The difference between energy cooperatives and crowdfunding consists in the way is structured. An energy cooperative is an organization which focus on raising funds for its own projects. On the other hand, crowdfunding platforms (Citizenenergy, 2016) may have multiple different projects in different countries. Currently, cooperatives can make use of crowdfunding platforms. In addition to the two mentioned above way to finance projects, there is another type of finance project classified as hybrid, called playfunding, a platform that puts in touch crowd with RES projects and sponsors. The crowd uses a web platform and watch the advertisement of a sponsor. The sponsor provides the RES project with money for each view or package of views.

Another form of fund for sustainable energy projects is donation, which is based crowdfunding, but you make an investment without any expectation of a financial or material return. At the other pol, is reward based crowdfunding, when your return for investment comes in the form of electricity supplied or a discount on electricity rates.

In the past years, there has been seen a significant increase of citizens' investment and engagement with renewable energy. However, there is still more to do and place for improvement. There is an urgent need to align national crowdfunding legislation, promote and share best practices both an national, and regional level. In European Union, currently there are a wide variety of national legislation which apply to crowdfunding across Member States (Citizenenergy, 2017). This depends on the model chosen by platform owners, investment amounts and other factors. In the future new formulas will exist.

CHALLENGES ON IMPLEMENTATION OF BUSINESS MODELS FOR ENERGY APPLICATIONS ON ISLANDS

In general, renewable energy projects face a number of challenges as every business venture. Renewable energy projects such as wind, solar and biomass although considered as mature technologies still encounter difficulties associated with risks stemming from the aforementioned characteristics as well as specific details of the project as defined by the geographical and techno-economic specifications of the project. Business models are called to address technical and technologies challenges, administrative, market and financial challenges during RES projects implementation (Pasicko, 2016b).

Although remote areas such as islands could demonstrate as living labs for applying both mature and developing technologies, they face several complications originating from their 'isolated' nature. Remoteness can be defined by different attributes infrastructures, geographic and economic (Vallve, 2013) affecting the realisation of renewable energy projects. Due to the infrastructural, geographic and economic remoteness, the islands face security of supply risks throughout the year, high electricity prices, as they usually use subsidized diesel power engines and high emission intensity factors. In addition, islands lack of local tailor-made regulation codes, which provide room for cooperative ventures upon the local communities. In contrast, they lack market opening and liberalization policies, which lead the public corporations to preserve their vertical roles as usually energy producers and operators with no room for private entities to penetrate the market. Frequently, projects on islands are considered either too small or too risky due to grid instabilities, and low demand levels with no investment interest and lack of financing resources. These handicaps should be tackled effectively through tailor made business models which could address the candidate projects on a local, tailor-made-level.

Business models could facilitate as platforms for RES development and offer to the potential off-takers a profit from renewable energy on islands. Due to the fact that several stakeholders located on islands do not have the required knowhow in electricity generation, prospects for business ventures arise for external investors or for business models incorporating a rental / leasing structure.

Different factors could impact the optimal electricity generation implementation on islands as described in Table 1. These range from external factors such as sun and wind conditions, grid-availability or grid stability to those related to the island community, e.g. environmental sustainability approaches, accessibility of capital and whether or not the island state is directly in charge of the energy generation. Other components are maybe related to the island off-takers. Amongst these are load-profiles, availability of space straightforwardly next to the primary off-takers and current electricity

Table 1. Factors impacting the optimal electricity generation implementation on islands

Technical	Economic	Geographical	Regulatory/Social
Renewable energy potential (Solar, wind, geothermal, biomass etc)	Availability of capital (communities, stakeholders, private parties, companies, crow funding etc)	Distance from other islands/mainland	Operational Framework (Code) & Market Structure
Interconnectivity	Incentives	Terrain	Unbundling/Opening/ Liberalization
Grid-stability/ availability	Availability of financing	Areas/Zones configuration	Policies (environmental, competition, subsidization etc)
Load Profile		Space available	Social Acceptance

generation. These unique features describing a project lead to the adoption of different business models.

A sample of business model appropriate for energy applications on islands are described as follows (The Energy Sustainable Consulting, no date):

- **Self-Consumption (Plant Ownership):** The power plant is located on-site and a particular island end-user consumes the electricity. Another alternative of this BM would be integrating the "micro-grid" concept while selling the electricity surplus to it or to adjacent consumers
- **Power Purchase Agreement (Standard PPA):** The island off-taker/utility buys the electricity at a predefined cost from an Independent Power Provider (IPP) either off-grid or grid-connected. This IPP can operate either a conventional or a renewable energy power plant. Furthermore, in cases of an open market structure, flexible mechanisms such as link to diesel price or spot market price might be required.
- **Synthetic PPA:** The IPP sells at market price, participants in the electricity market declare a guaranteed price on the day ahead market usually based on their marginal costs and they profit from certain deviations as configured by the electricity generation mix merit order. This business model demonstrates a grid-connection and a working spot/day ahead market, which usually is not connected in the tremendous larger part of islands which work based on an autonomous basis.
- **Joint Venture (Co-ownership):** The island off-taker (local utility, end consumer) and a third-party investor (private companies, local community etc) establish a joint venture. This entity acts in the same way as an IPP and supplies power to *the island off-taker under an PPA energy.*
- **Leasing or Rental Agreements:** Island off-takers do not invest themselves they pay a leasing rate while they operate the renewable energy plant and consume or sell the energy surplus that the power plant generates.

The aforementioned Business Models for RES development on islands are illustrated in Figure 1. The schematic representation of it shows the links between parameters affecting the form of business models for RES projects linked with the different ownership structures, technologies, business models and other attributes of the project.

Figure 1. Decision Schematic Diagram for Business Models on Islands
Adapted after: The Energy Sustainable Consulting, no date.

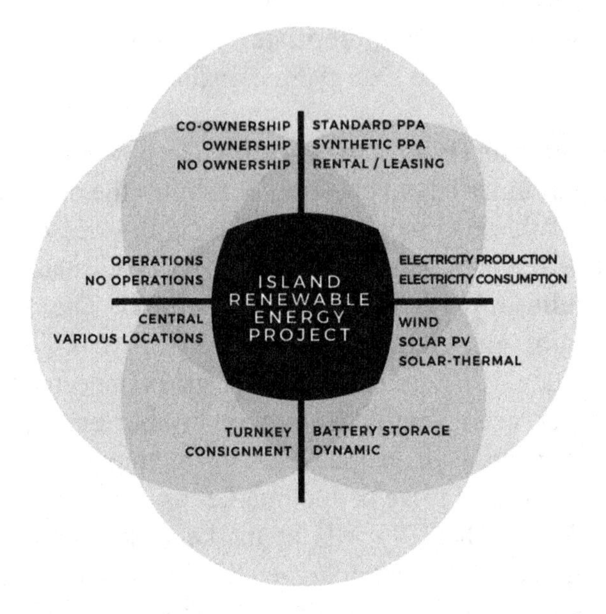

Figure 2. Stakeholders participating in Renewable Energy Projects on Islands and Remote areas
Adapted after Vallve, 2013.

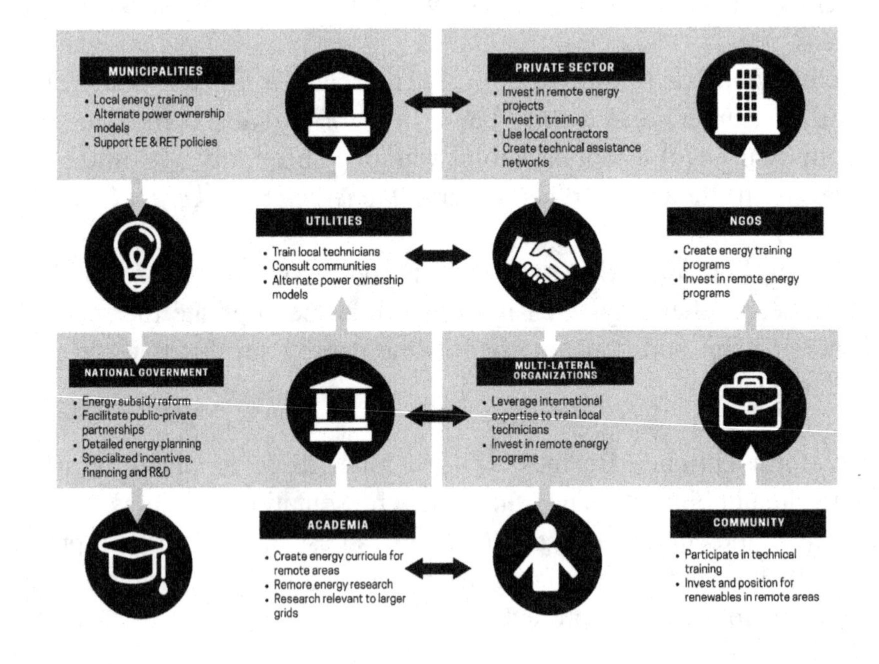

The main actors contributing in the life cycle of a renewable energy plant (plan, development, financing, construction, operation and decommission) on an island are summarized in Figure 2. Due to the relatively small size of the islands, the stakeholders can have different roles within the project (developer, operator, investor, owner etc) while more flexibility is added on

Table 2. Examples of different ownership structures on islands

Island	Project	Ownership
Ramea (Canada)	Wind Farm & hydrogen system	Partnership between 1) Frontier Power Systems and 2) Nalcor
Faroe Islands	Wind Projects, Solar Thermal and hydrogen system	1)Municipal utility 2)Community company 3)Community non-profit
Isle of Eigg (UK)	Mini-grid	Community-owned utility
Floreana (Galapagos)	PV/diesel/battery -2 diesel generation sets	1)Local authority 2) Local Utility
Bonaire (Caribbean)	Wind/Hydro/storage hybrid	Private independent power producer
Miyakojima (Japan)	PV/battery/smart grid	Utility
Reunion Island (France)	PV in different sizes, sola thermal, wind, ocean, hydro, algae, EVs	Different forms of partnerships, public, private and hybrid
El Hierro (Spain)	Wind/hydro pump/diesel	Public/private partnership

Valve, 2013.

Table 3. Financing structures for renewable energy projects on islands

Island	Project	Financing Structure
Faroe Islands	2.13 MW wind 1.98 MW wind 220 KW wind for electrical heat Solar thermal 200 m2, 10 kW wind & hydrogen storage	Utility Financed Private Financing 86% grant 14% community equity 100% grant to date
Floreana	PV/diesel/battery hybrid 2* 68 KW diesel gensets	International donors, national/local government & users National government & internation donours
Bonaire	Wind/Hydro/storage hybrid	100% private project finance 80/20 debt/ equity
El Hierro	Wind/hydro pump/diesel	10% private fiannce 35% public finance 55% grant

Valve, 2013.

the community side to act as either as consumer or as producer.

A number of case studies of different ownership status on islands, which have applied different business models is presented in Table 2.

Examples of the financing structure incorporated in the different business models are provided in Table 3 for three of the aforementioned projects.

RISKS FOR IMPLEMENTATION OF BUSINESS MODELS ON ISLANDS

Financing projects on remote areas such as islands embeds higher risks compared to those located in densely populated and well-interconnected areas. Usually the cost of capital is higher, and there are no sufficient, bespoke financial instruments to support such business models. The main risks are associated with the timing and the scheduling of the project stages, which usually can be affected from geographical remoteness. Furthermore, hazardous areas due to extreme weather events, earthquakes and so on has an impact in the configuration of the details of business plan and in accumulating capital equity. Other parameters are related with the regulation framework and the mechanism supporting renewable energy, electricity prices, duration of the power purchase agreement etc. Other parameters are related with the sufficient weather data collection and analysis as well as with the performance of the project as indicated in the energy study. Last but not least, the financial strength of the counterparties participating in the project as well as the general economic conditions of the country and the local region could affect significantly the access to financing and the cost of capital.

Ways of mitigating risks could be the assurance of the technical viability of the project through a series of certified measurements and careful consideration of all the technical and procedural details of the project. Furthermore, a Credit Line which can provide a line to the local banks which they could provide financing to the remote communities, enhancing the role of guarantees participating at the project. Loan funds from the government or utility funds with low interest rates and long payback periods as well as other incentives from the government such as grants and partial subsidization could boost renewable energy development on islands while configuring feasible business models for the islands Moreland Energy Foundation, Net Balance and Green Spark Consulting, 2011; Vallve, 2013).

REACTION IN TERMS OF STRATEGIES AND POLICY INITIATIVES FOR SCALING UP RENEWABLE ENERGY DEVELOPMENT

One of the challenge for RE deployment in islands nations is capacity for planning the development of the energy sector, set the right policy framework for investment. The critical role of renewables for sustainable development strategies has been highlighted in a number of declarations. Such examples are the declarations of the Canary Islands (1998), Palma de Majorca (1999), Azores (2000), Cagliary (2001), Chania (Crete-2001) and so on. Partnership was strongly highlighted as a key to success in terms of progress. Numerous examples of declarations (such as Declaration of Chania June 2001), campaigns (Island 2010 Campaign), summit (Island Solar Summit 2000) all highlight the importance of different players (Marin et al., 2015). Chania Declaration also highlighted that building 100% RES Island Communities ought to incorporate all divisions related to transport, water supply and management. We reviewed policy and strategic planning initiatives worldwide in islands. Example of existing policy commitments of renewable energy support in islands is provided in Table 4.

A chronological list of policy initiatives and strategic action planning for islands is provided in Table 5 below, followed by a discussion. Sectors gathering interest for paving a sustainable future of islands are mentioned (Table 5 column 5). This will help the author to decide which primary topics should considered to discuss in this book.

Tokelau National Energy Policy and strategic action planning (NEPSAP) (O'Brien, 2004) is published by the government of Tokelau aiming to transform the island to an energy depended system while applying energy efficiency techniques and accelerate renewable energy planning. The policy includes measures for all the sectors including transport, electricity while addressing carefully matters related to fuel dependency.

The General Electricity Act published in 2007 (NREL Dominican Republic, 2015). in Dominican Republic supports energy savings while allowing CNE

Table 4. Existing Policy Commitments in Caribbean

	Feed-in Tariff	Net Metering/ Billing	Renewables Portfolio Standard/ Quota	Tax Credits	Tax Reduction/ Exemption	Public Loans/Grants
Antigua and Barbuda		D		D	D	
Bahamas	ID	ID			D	
Barbados	ID	D	ID	D	D	D
Dominica		D			D	
Dominican Republic		D		D	D	D
Haiti						ID
Jamaica		D			D	D
Saint Kitts and Nevis				D	D	
Saint Lucia		D			D	
Trinidad and Tobago	D			D	D	

Adapted from Timilsina & Shah, 2016.
ID- in development
D – developed

to act for matters related to energy efficiency, renewable energy, projects, programs, policies and standards. Furthermore, Law 57-07, the law for Renewable Sources of Energy Incentives and Its Special Regimes proposes several initiatives and acts to accelerate RES development in the Dominican Republic. Specifically, law 57-07 offers a number of tax incentives for energy conservation and renewable energy such as a 10-year feed-in tariff (FiT) for connected to the grid renewable energy projects.

The Kiribati Development Plan (KDP) established for the years 2008-2011 (Government of the Republic of Kiribati,2009) put the foundation for the necessary improvements that the island required in the energy sector based on a number of region scenarios. The main target was to provide a framework in which investors and stakeholders will count on designing their future investments and actions in the energy sector. This plan will be maintained and reviewed whenever is considered appropriate. The local government tried also to reflect the role of energy in achieving an overall quality of life improvement on the island.

The US Virgin islands passed the Act 7075 (2009) (NREL U.S. Virgin Islands, 2015) which targets on eliminating fuel imports on the islands while

Table 5. Policy initiatives and strategic action planning for islands

N	Policy and Strategic Planning	Year of Release / Establishment	Region Coverage	Topics Coverage	Source
1	TOKELAU NATIONAL ENERGY POLICY AND STRATEGIC ACTION PLANNING (NEPSAP)	2004	Tokelau	Energy planning and regulation Electricity Energy efficiency Energy independence Transport Energy and the environment	O'Brien, 2004
2	GENERAL ELECTRICITY ACT OF 2007 & LAW 57-07	2007	Dominican Republic	Energy efficiency Renewable Energy Market	NREL Dominican Republic, 2015
3	KIRIBATI NATIONAL ENERGY POLICY (KNEP)	2009	Kiribati	Human and Institutional Resource Development Energy Security Economic Growth and Improvement of Livelihoods Sustainability, Equity Environment Stakeholder Participation Governance, Compatibility Policy planning and coordination for Power Islands and Rural Electrification Efficiency and Conservation Transport Renewable Energy	Government of the Republic of Kiribati, 2009
4	U.S. VIRGIN ISLANDS Act 7075	2009	US Virgin Islands	Renewable Energy Energy Efficiency Heating	NREL U.S. Virgin Islands, 2015.
5	JAMAICA ENERGY POLICY	2009	Jamaica	Renewable Energy Energy Efficiency	NREL Jamaica, 2015
6	ENERGY POLICY IN CURACAO	2009	Curacao	Targets Renewable Energy	NREL Curacao, 2015
7	ACT 82 by ENERGY PUBLIC POLICY OFFICE	2010	Puerto Rico	Renewable Energy	NREL Puerto Rico, 2015
8	SUSTAINABLE ENERGY FRAMEWORK OF BARBADOS	2010	Barbados Islands	Interconnection Standards Market Energy Efficiency	NREL Barbados, 2015
9	NATIONAL ENERGY POLICY OF PALAU	2010	Palau	Targets Energy Efficiency Renewable Energy	NREL Palau, 2015
10	MALDIVES NATIONAL ENERGY POLICY AND STRATEGY	2010	Maldives	Sustainable development Energy Security Energy efficiency Energy resources and resource management Energy equity	Maldives Energy Authority (N/A)
11	ENERGY POLICY IN CURACAO (2)	2011	Curacao	Electricity Market	NREL Curacao, 2015

continued on following page

Table 5. Continued

N	Policy and Strategic Planning	Year of Release / Establishment	Region Coverage	Topics Coverage	Source
12	NATIONAL ENERGY POLICY MICRONESIA	2012	Federal States of Micronesia	Renewable Energy Energy efficiency	NREL, Micronesia 2015
13	STATE OF HAWAII – ENERGY POLICY DIRECTIVES	2013	Hawaii	Interconnections Infrastructure Renewable Energy Waste and water management Environment	Hawaii state energy office (N/A)
14	ACT 57 by ENERGY PUBLIC POLICY OFFICE	2014	Puerto Rico	Sustainability Energy efficiency Renewable energy Cost efficiency	NREL Puerto Rico, 2015
15	NATIONAL ENERGY POLICY OF BARBADOS	2014	Barbados Islands	Energy Efficiency Renewable Energy Energy costs	NREL Barbados, 2015
16	SOLOMON ISLANDS NATIONAL ENERGY POLICY AND STRATEGIC PLAN	2014	Solomon Islands	Health Economy Development Environment Ecosystem Energy	Ministry of Mines, Energy and Rural Electrification, 2014
17	LEGAL AND REGULATORY FRAMEWORK OF THE ENERGY SECTOR OF THE MALDIVES	2016	Maldives	Energy Renewable Energy Infrastructure Market Power Efficiency Costs	Maldives Energy Authority (N/A)
18	NATIONAL ENERGY POLICY 2017 - 2037 CAYMAN ISLANDS	2016	Cayman Islands	Renewable energy Energy efficiency and conservation Energy security Sustainable Environment	Cayman Islands Government, 2013
19	GOVERNMENT OF THE VIRGIN ISLANDS ENERGY POLICY	2016	British Virgin Islands	Energy Environment Consumer Awareness Job and Industry Development Energy Security and reliability	Government of the Virgin Islands, 2016

increasing alternative forms of energy. The measures included in the act propose:

- Increase of renewable energy penetration at 30% in the electricity mix
- Net metering for distributed generation
- New constructions are imposed to install energy-efficient solar water heating systems by 2020
- 32 million euros were extracted from the Recovery Act to finance projects and initiatives for energy efficiency and distributed renewable energy

Jamaica established a target of 50% energy efficiency from 2015 to 2030 (NREL, Jamaica, 2015) as part of its national policy in 2009 its initial inclusive long-term energy plan. The policy established several targets by 2030 related to renewable electricity generation, energy efficiency, and greenhouse gas emissions reduction.

The state of Curacao published the energy policy strategy which presents directions and governing principles for improving key performance indicators in the energy sector. One of the main targets included is the reduction of energy consumption by 40% by 2020 (NREL, Curacao, 2015). In addition, the policy supprost energy storage in parallel with wind development to supply 100% of the island's energy needs.

Act 82 by Energy Public Policy Office in Puerto Rico was voted in 2010 (NREL Puerto Rico, 2015), which proposes a renewable energy portfolio for Puerto Rico to be differentiated proposing 20% RES integration in the electricity sector by 2035.

The Sustainable Energy Framework for the Barbados islands (NREL Barbados, 2015) was published in June 2010 with the assistance of the Inter-American Development Bank, and investigates future energy scenarios and makes a series of policy recommendations to accelerate sustainable energy development.

In 2010, The Republic of Palau endorsed its National Energy Policy (NREL Palau, 2015). A guiding document for implementation of this policy An Energy Sector Strategic Action Plan has been defined. The policy refers to institutional arrangements; increased energy efficiency and renewable energy; ensure security, reliability, and efficiency of the electricity supply. The NEP set targets to reduce national energy consumption 30% by 2020 and produce a minimum of 20% of total energy from renewable sources by 2020.

Principles of the Energy Policy and Strategy for Maldives (Ministry of Housing and environment, 2010) include:

1. Reduce overreliance of the energy sector
2. Improve energy efficiency and conservation of energy use
3. Encourage the adoption of low-carbon innovations in generation
4. Exploit local energy resources and renewable technologies
5. Engage private sector participation in the development of the energy sector, energy services and quality assurance mechanisms
6. Ensure energy equity through social protection mechanisms or safety nets for vulnerable groups of population

In 2011, the government of Curacao published a policy document for the regulation of the electricity sector, which explained on the thoughts sketched out in the 2009 policy approach (NREL Curacao, 2015). This later document encouraged competition by licensing independent power producers and lowering cost in the generation sector. It moreover empowered the elimination of import taxes on renewable energy generation equipment and the creation of a tax credit for the installation of those systems.

The National Eergy Policy Micronesia (NREL Micronesia, 2015) document is divided into two volumes: (1) renewable energy goals and energy efficiency goals; (2) national and state level energy action plans to achieve the overall targets. The first volume includes an increase of share of electricity generated from renewable sources to 30% by 2020 while also increasing energy efficiency by 50%. It too diagrams a few other targets, such as raising the average energy generation efficiency of conventional generating units 20% by 2015. Another one is increasing the rural electrification rate to 90% by 2020. The second volume focuses on energy efficiency in public facilities, energy standards for all buildings and energy awareness campaigns.

State of Hawaii – through its energy policy directives aims to diverse resources such as solar, wind, hydro, bioenergy, geothermal, and energy efficiency (Hawaii State Energy Office). Interconnections make use of best resources, which could help improve overall system efficiency and lower electricity rates across islands. The State Administration is determined to achieve its goal of 100 percent renewable energy generation by 2045. Hawaii is seen as an ideal test bed for new energy solutions.

In 2014, Act 57 by energy public policy office in Puerto Rico (NREL Puerto Rico, 2015) codified noteworthy and broad changes to the electricity sector, laying the establishment for a more sustainable and cost-effective energy system in Puerto Rico.

The National Energy Policy of Barbados (NREL Barbados, 2015) was published in 2006 and characterized a wide set of standards, including maximizing efficiency of energy use at all stages of utilization, decreasing fossil fuel utilization, increasing use of renewable energy .

The Solomon Islands National Energy policy and strategic plan has a number of objectives consisting in the following (Ministry of Mines, Energy and Rural Electrification, 2014): to reduce poverty and give more prominent benefits to Solomon islanders, to provide support to the vulnerable; to ensure access to health care, to education, to increase the rate of economic growth and equitably distribute the benefits of employment amongst Solomon Islanders, to build and upgrade physical infrastructure and utilities to ensure all islanders have access to essential services; to protect the environment and ecosystems and islanders from natural disasters; to improve governance at all levels and strengthen links between them.

The legal and regulatory framework of the energy sector of the Maldives (2016) focuses on generation, distribution and supply licencing regulation, Net Metering regulation (Maldives Energy Authority, N/A). The objectives are:

- Improving the safety, quality and reliability of the power service industry
- Improving the economic operation of power systems
- Protection for investors, service provider and consumers
- Incorporating support for renewable energy and energy efficiency as per energy policy Complementary to the SREP to provide an enabling environment for power sector investments

The ultimate goal of the National Energy policy for 2017-2037 of Cayman Islands (2016) is to contribute to the global reduction of greenhouse gas emissions through the use of environmentally friendlier sources of energy . We have to accept that, given the small size of this region, the contribution will be minimal when compared to emissions of much larger countries. As a small Island nation that is extremely vulnerable to the impact of climate change it is important to progress, and in doing so we, as a country and individually, can also benefit economically from being less reliant on fossil fuels.

The Energy Policy of the Government of the Virgin Islands Energy Policy (2016) address the following (Government of the Virgin Islands, 2016):

- Energy Policy Goals to incorporate into the policy. Goals are measurable objectives which establish how the Government will plan for future energy activities.
- Energy Targets measure desired outcomes for the Energy Policy Goals.
- Guiding principles to establish criteria for instituting energy targets, strategies, and action plans.
- Actions as a set of activities within the context of an overarching strategy which supports the goals of the Energy Policy.

Energy policies and strategic action plans supported by most of the Pacific Islands address primary renewable energy, amongst others. They incorporate explanations advancing the deployment of RE and expanding its share of the energy mix as well as empowering the reduction of greenhouse gas (GHG) emissions. The data displayed infers that islands as Fiji, Papua New Guinea, Samoa and Vanuatu have a noteworthy share of RE for electricity generation, generally from hydropower. Islands are progressively embracing national energy policies and RE targets, be that as it may there is a requirement to support these endeavors with a clear guide and roadmap and a nitty gritty usage plan counting the assignment of financial resources required to accomplish the RE target.

The most widely adopted policy is Feed-in tariffs (FITs). This provides a fixed, guaranteed price per unit of energy produced and sold into the grid. Only few islands have adopted this. Net metering or net billing schemes have been introduced and tested in some islands. Electricity consumers will have the ability to generate their own electricity and to feed it into the grid. The customer is billed or credited/paid based on the ratio of power consume to power generated. In Jamaica, those consumers who generate their own electricity, the electricity generation is deducted from their bill. In the Cook Islands and Palau, net-metering policies have been adopted, allowing private investors to support grid-connected solar.

FUTURE RESEARCH DIRECTIONS

Around the world, national energy policies have been developed, practical strategic action plans to implement the policies have been adopted, commonsense instruments to execute the activity plans have been created. In order to encourage the increase of renewable energy and simulate low carbon economic growth appropriate policies mixes are needed. Generally, countries that encourage renewable energy have a long goal vision, clear policies and measures to support the vision and well defined governance structures to implement them. These vary by region targets.

For example in the islands which are members of the Pacific Islands Forum Secretariat (Cook Islands, Federal State of Micronesia, Fiji, Kiribati, Nauru, Niue, Palau, Papua New Guinea, Republic of the Marshall Islands, Samoa, The Solomon Islands, Tonga, Tuvalu and Vanuatu) the Implementation Plan of the Framework for Action on Energy Security in Pacific (FAESP) the joint long-term goal is to increase investment in the RE technologies and four priorities have been identified: resource assessment; investment in RE; capacity development; increase in the share of RE in the energy mix (FAESP, 2010). Almost of these islands specified policies and goals to increase their use of renewable energy for power generation and reducing fuel imports (Timilsina & Shah, 2016). The problem is the long history of monopoly control. Some of them allowed changes to laws.

Despite the fact that numerous islands have presented and introduced targets plans to advance RE deployment, the policy design and execution it is frequently behind. Furthermore, there are still significant remaining things to be done. For example more emphasis needs to be given on sustainability of energy investments, business models and financing modalities that provide ownership and contribution of consumers and other stakeholders. This requires a strong role of commercial banks and lending agencies. Testing business models in practice and putting solutions into effect should be at the top of the pyramid of priorities for islands.

CONCLUSION

It has been concluded that in renewable energy projects, ownership business models center in specialized complexity, economies of scale, capital costs, and financing perspectives based on its own characteristics. It has many favourable features, including ability to provide power local communities, create jobs. However, business models include several decisive financing, service, and monitoring characteristics. They should be dynamic, while being adjusted to the special conditions, features and risks of the given project. In renewable energy projects, ownership business models center on the specialized complexity, economies of scale, capital costs, and financing perspectives.

It has also been concluded that the public-private partnership (PPP) is usually the optimum business model option for medium to large scale or grid connected renewable energy projects, and is usually applied with a structure, of a built-own-operate-transfer or multiparty ownership.

REFERENCES

Asian Development Bank (ADB). (2015). *Business models to realize the potential of renewable energy and energy efficiency in the Greater Mekong Subregion*. Mandaluyong City, Philippines: Asian Development Bank. Retrieved from: https://www.adb.org/sites/default/files/publication/161889/business-models-renewable-energy-gms.pdf

Bakker, S., & Trip, J. J. (2013). Policy options to support the adoption of electric vehicles in the urban environment. *Transportation Research Part D, Transport and Environment*, *25*, 18–23. doi:10.1016/j.trd.2013.07.005

Bauwens, T., Gotchev, B., & Holstenkamp, L. (2016). What Drives the Development of Community Energy in Europe? The Case of Wind Power Cooperatives. *Energy Research & Social Science*, *13*, 136–147. doi:10.1016/j.erss.2015.12.016

Beeton, D., & Meyer, G. (2015). *Electric Vehicle Business Models, Series Title Lecture Notes in Mobility*. Publisher Springer International Publishing.

Bolton, R., & Hannon, M. (2016). Governing sustainability transitions through business model innovation: Towards a systems understanding. *Research Policy*, *45*(9), 1731–1742. doi:10.1016/j.respol.2016.05.003

Centre For Sustainable Energy. (2009). *Delivering community benefits from wind energy development: A tool kit. Report for the Renewables Advisory Board and DTI*. Department of Trade and Industry. Retrieved from: https://www.cse.org.uk/downloads/toolkits/community-energy/planning/renewables/delivering-community-benefits-from-wind-energy-tookit.pdf

Citizenergy. (2016). *Fostering crowd-based investments in renewables and energy efficiency*. Retrieved from: https://citizenergy.eu/themes/citizenergynew/assets/documents/2016_06_Position_Paper_Fostering_Crowd_Based_Investment.pdf

Citizenergy. (2017). *The European platform for citizen investment in renewable energy*. Retrieved from https://citizenergy.eu/themes/citizenergynew/assets/documents/2017_02%20Legal%20Review%20(Update).pdf

European Commission. (2009). *Regulation (Ec) No 443/2009 Of The European Parliament And Of The Council of 23 April 2009. setting emission performance standards for new passenger cars as part of the Community's integrated approach to reduce CO2 emissions from light-duty vehicles*. Retrieved from: http://eur-lex.europa.eu/legal- content/EN/TXT/?uri=celex:32009R0443

Hall, S., & Roelich, K. (2016). Business model innovation in electricity supply markets: The role of complex value in the United Kingdom. *Energy Policy*, *92*, 286–298. doi:10.1016/j.enpol.2016.02.019

Leaney, V., Jenkins, D., Rowlands, A., Gwilliam, R., & Smith, D. (2001). Local and community ownership of renewable energy power production: Examples of wind turbine projects. *Wind Engineering*, *25*(4), 215–226. doi:10.1260/0309524011496033

Moreland Energy Foundation, Net Balance, and Green Spark Consulting. (2011). *Business models for enabling sustainable precincts*. Research Report Prepared for Sustainability Victoria. Retrieved from: https://www.parliament.vic.gov.au/images/stories/committees/osisdv/Growing_the_Suburbs/Submissions/Sub_32_Moreland_Energy_Foundation_7.12.2011_Growing_the_Suburbs_OSISDC_Attachment_A.pdf

Palan, R., Murphy, R., & Chavagneux, C. (2010). *Tax Havens How Globalization Really Works* (1st ed.). Ithaca, NY: Cornell University Press.

Pasicko, R. (2016a). *Crowfunding for Renewable Energy Systems*. Brussels: 2nd Summit of OCT Energy Ministers. Retrieved from: http://www.octassociation.org/2nd-summit-of-oct-energy-ministers

Pasicko, R. (2016b). *Innovative financing and business models for RES*. Retrieved from: http://www.irena.org/EventDocs/UNDP%20Croatia,%20Innovative%20financing%20and%20business%20models%20for%20RES.pdf

Perdiguero, J., & Jiménez, J. L. (2012). Policy options for the promotion of electric vehicles: a review. *Research Institute of Applied Economics. Working paper, 20*, 208.

Rijpens, J., Riutort, S., & Huybrechts, B. (2013). *Report on REScoop Business Models*. Available at: http://www.rescoop.eu/sites/default/files/project-resources/report_on_existing_business_models_deliverable_3.2.pdf

Roberts, S., Letcher, R., & Redgrove, Z. (2007). Mobilising individual behavioural change through community initiatives: Lessons for tackling climate change. *The Energy Review Study*, (298740), 1–7.

Roehrich, J. K., Lewis, M. A., & George, G. (2014). Are public-private partnerships a healthy option? A systematic literature review. *Social Science and Medicine. Elsevier Ltd*, *113*, 110–119. doi:10.1016/j.socscimed.2014.03.037 PMID:24861412

Rogers, J. C., Simmons, E. A., Convery, I., & Weatherall, A. (2012). Social impacts of community renewable energy projects: Findings from a woodfuel case study. *Energy Policy*, *42*, 239–247. doi:10.1016/j.enpol.2011.11.081

The Energy Sustainable Consulting. (n.d.). *Business models for renewable energy applications on islands*. Retrieved from: https://www.th-energy.net/english/platform-renewable-energy-on-islands/business-models/

Vallve, X. (2013). *Renewable Energies for Remote Areas and Islands (Remote)*. Microgrids Symposium, Santiago, Chile. Retrieved from: https://building-microgrid.lbl.gov/sites/all/files/santiago_vallve.pdf

Viardot, E. (2013). The role of cooperatives in overcoming the barriers to adoption of renewable energy. *Energy Policy*, *63*(C), 756–764. doi:10.1016/j.enpol.2013.08.034

Walker, G. P. (2008). What are the barriers and incentives for community-owned means of energy production and use. *Energy Policy*, *36*(12), 4401–4405. doi:10.1016/j.enpol.2008.09.032

Wirth, S. (2014). Communities matter: Institutional preconditions for community renewable energy. *Energy Policy*, *70*, 236–246. doi:10.1016/j.enpol.2014.03.021

Wizelius, T. (2014). *Windpower Ownership in Sweden: Business Models and Motives*. Routledge.

ADDITIONAL READING

European Commission. (2016) Air Pollutants from Transport, European Commission, Retrieved from: http://ec.europa.eu/environment/air/transport/road.htm

Heidrich, O., Rybski, D., Gudipudi, R., Floater, G., Costa, H., & Dawson, R. (2016) RAMSES- D1.3 Methods inventory for infrastructure assessment. Newcastle upon Tyne, UK. Retrieved from: http://www.ramses-cities.eu/results

Leonardi, P. M. (2010). From road to lab to math the co-evolution of technological, regulatory, and organizational innovations for automotive crash testing. *Social Studies of Science*, *40*(2), 243–274. doi:10.1177/0306312709346079 PMID:20527322

Maingot, A. (1998). Laundering Drug Profits: Miami and Caribbean Tax Havens. *Journal of Interamerican Studies and World Affairs*, *30*(2/3), 167–187. doi:10.2307/165985

Rogers, R. (1991). TheVehicle Helping People Survive and Avoid Crashes. *ITE Journal.*

San Román, T. G., Momber, I., Abbad, M. R., & Miralles, A. S. (2011). Regulatory framework and business models for charging plug-in electric vehicles: Infrastructure, agents, and commercial relationships. *Energy Policy*, *39*(10), 6360–6375. doi:10.1016/j.enpol.2011.07.037

Tomi, J., & Kempton, W. (2007). Using fleets of electric-drive vehicles for grid support. *Journal of Power Sources*, *168*(2), 459–468. doi:10.1016/j.jpowsour.2007.03.010

Wells, P. E., & Nieuwenhuis, P. A. H. F. (2007). The all-steel body as a cornerstone to the foundations of the mass production car industry. *Industrial and Corporate Change*, *16*(2), 183–211. doi:10.1093/icc/dtm001

Zapata, C. and Nieuwenhuis, P. (2010) Exploring innovation in the automotive industry. *Journal of Cleaner Production* 18(1), pp. 14-20. (.10.1016/j.jclepro.2009.09.009

Chapter 5
Indicators, Modelling, and Visualization of Islands

ABSTRACT

In this chapter, the author presents a discussion on the range and usefulness of energy models in an island context, pros and cons of the various methodologies, and selection criteria that could guide a proper model choice for further implementation. The analysis has been done considering a clustering approach. Following this, the author presents a "toy" model called ISLA (sustainable islands), which has been developed for students to evaluate energy systems in islands. The model has been tested on numerous islands so far. For demonstration purpose, Crete island has been considered as a case study to capture the model flexibility and adaptability features. The rationale of using the model is discussed, together with the data needed, the validity and usefulness of outcomes produced, and the way such tools can guide policy making. Comparing the toy model with a sophisticated approach demonstrates some interesting advantages offered by this methodology and visualization.

ISLAND INDICATORS AND CRITERIA FOR CLASSIFICATION

Islands are all of sort of forms, shape, size and have diverse characteristics. Indicators could provide us understanding on the nature of islands and identify islands with particular characteristics. Measuring the isolation of an island

DOI: 10.4018/978-1-5225-6002-9.ch005

from other human inhabited places, can give us an indication of isolation. Or measuring the percentage of the land area of an island above sea level can give us an indication of the risk of the sea level rise. Measuring the level of economic development and impact on the environment considering income per capita and gross domestic product can provide an indication of the wealth of the islands. However, there are no statistics to allow the calculations to be applied for islands.

The benchmarking study could be done based on the cluster methodology. For example let's assume we have a group of seven islands with similar climatic conditions, but different socio-economic characteristics. First step is to analyse the socio-economic characteristics and organize them in a number of clusters. Let's assume we end up with four clusters. In this case, the seven islands could belong to the same climatic cluster, but to the four different socio-economic clusters. The primary objective is to have the ability to design renewable energy systems for islands with similar conditions. Although islands have commonalities, when it comes to policy making, careful treat must be provided. The following list provides a template for classifying islands around the world to assist in policy making and planning (Sheldon, 2005):

- Climate Characteristics (cold, temperate or tropical).
- Proximity to the mainland, since islands that are more remote and distant are challenged to cope with their remoteness and they face complications in accessibility and connectivity
- Size and impact on increase of number of tourisms during seasonal term
- Proximity to other islands
- The Governance of the island plays a role in the support of sustainable development plans. Depending on if the island is an autonomous state or they are part of larger countries and follow the same national or regional policies could impact future plans based on the applied national strategies.
- Population levels/Economic Growth is significant in designing sustainable policies**.** Usually in such islands sustainable environmental practices in tourism can found prosperous ground even inside the local community while in larger islands centralized policies are required applicable to the different sub-sector of tourism

Table 1. Energy models applied for islands case studies

Model	Link	References
EnergyPLAN	http://www.energyplan.eu	Connolly et al. (2011)
H_2RES	http://h2res.fsb.hr	Chen et al. (2007)
MARKAL/ TIMES	http://www.iea-etsap.org/web/Markal.asp	Das and Ahlgren (2007)
ENPEP-BALANCE	http://ceeesa.es.anl.gov/news/EnpepwinApps.html	Centre for Energy, Environmental and System Analysis
Invert	http://www.invert.at	Tsioliaridou et al. (2006)
LEAP	http://www.energycommunity.org	Giatrakos et al (2009)
SimREN	http://www.energyplan.eu/othertools/national/simren/	Lehmann (2003)
UniSyD3.0	http://www.energyplan.eu/othertools/national/unisyd3-0/	Leaver et al (2012)
WILMAR Planning Tool	http://www.wilmar.risoe.dk/project_description.htm	Mebom et al (2007)
HOMER	http://www.homerenergy.com/index.html	Demiroren and Yilmaz (2010)
RETScreen	http://www.nrcan.gc.ca/energy/software-tools/7465	Bakos and Soursos (2002)
MESSAGE	http://www.iiasa.ac.at/web/home/research/researchPrograms/Energy/MESSAGE.en.html	International Atomic Energy Agency (2008)
TRNSYS	http://www.trnsys.com	Kalogirou (2001)

- Homogeneity of the population and the socio-cultural sustainability of island goals affect the resilience of the locals to large tourism waves.

ISLANDS STUDIES AND MODELS

We have reviewed some of the existing energy models and which of them have been applied to islands. This section seeks to compile together the information (Table1). The list is not exhaustive.

The following key areas of practice have been identified: renewable energy integration and climate change, technology design and policy assessment, energy and transportation modeling. Despite the diversity of models, the approaches to islands energy system modeling share the following common challenges: data quality and uncertainty, model integration, and policy relevance. A key aspect is to what extent are modeling attempts limited by insufficient data.

Table 2. Medoid characteristics for islands studies

Category	Spatial	Temporal	Method
Technology	Technology	Monthly	Simulation
Energy demand	Buildings Transport Island	Annual, Monthly, Hourly	Simulation, Optimization
Whole Energy System design	Island	Static, Dynamic	Optimization
Policy analysis	Island	Static	Empirical
100% renewable energy system	Island	Dynamic	Simulation

To understand current practices, we review current literature and their applications to islands, to identify current practices. First, we identified a number of clusters within the data. This was done using a grid-based clustering approach based on a multiple-level granularity. We then construct a partition of the database into a set of k nonempty subsets, then partitioning around medoids in order to determine the attributes of each cluster. We determine four distinct categories (Table 2) with their allocated attributes: technology, energy demand, whole energy system design, policy assessment, and 100% renewable energy system

Each of these categories refers to different temporal resolutions. In the case of technology the temporal resolution vary from seconds (for EV studies) to hourly or daily (PV, hot water, cooling systems), while whole energy system design refers to whole island energy system using optimization techniques. Technologies refers to studies on operational analysis of wind power (Carta et al., 2003), small-scale biodiesel production (Skarlis et al., 2012), wind power pumped storage (Katsaprakakis et al., 2012) and so on. Whole energy system design studies refer to multi-objective optimization in energy systems (Koroneos et al., 2004), sustainable planning (Giatrakos et al., 2009). Policy analysis refers to policy and interventions for scalling up renewable energy development (Timilsina, et al., 2016).

I personally think if you wish to understand better how the energy system works the best is to try yourself to do an energy toy model. In the next section I explain how to construct such a toy model to simulate the whole energy system in an island nations and assess future scenarios. I say "island nations" because it can be applicable to any island size and type. To demonstrate this, the toy model has been applied for Crete. The next section provides a description of the toy model, and results for the case study.

A 'TOY' MODEL FOR SUSTAINABLE ISLANDS (ISLA)

To set up a common rule for the integration of renewable energy systems (RES) on any island is a difficult task. Resources shift and vary in a huge amount, as well as needs and island characteristics. Clearly, the approach for fueling with RES an island with 10,000 occupants is totally distinctive from one with half a million occupants or a country island.

In this section a simple energy simulation model for sustainable islands called ISLA is described. The aim is to analyze the future energy demand and supply scenarios for any island based on changes in historical trends, technology costs and performance, policies and given information on climate. The method can be applied to any island, even though input parameters need to be adjusted. The most challenging tasks are the assessment on regulation, storage and demand side response. Storage, because a variety of technologies should be included: batteries, hydropower storage, thermal storage etc. Also to achieve 100% energy solutions for an island, transport should also be studied. The Time Horizon of ISLA is 2050 and the temporal resolution 5 years. The architecture of the ISLA model is shown in Figure 1.

The first part of ISLA focuses on the demography of the island and simulates future population, average household size and number of households. The formulas used in the demographic section are as follows;

$$New\ Population = PreviousPopulation + \left(\left(GrowthScenario * PreviousPopulation\right) * TimeStep\right)$$

$$New\ Avg\ Household\ Size$$
$$= PreviousAvgHouseholdSize + \left(\left(GrowthScenario * PreviousAvgHouseholdSize\right) * TimeStep\right)$$

$$Number\ of\ Housholds = \frac{Population}{Avg\ Houseshold\ Size}$$

The second part of ISLA focuses on the energy demand of the sectors of the island namely; domestic, industrial, services and tourism. Service demands could include energy demand from education, health, military and retail sectors and tourism depending on island activities. The energy demand for each of the sectors are further divided into the types of energy used, i.e. electricity, heat, gas and oil categories. Furthermore, each of types of energy are split into their various end uses which varies from sector to sector. The ISLA model is very flexible in terms of input data, thus the energy and end-uses data used for simulation is dependent on the data availability, size, and economic and social activities of the island. In order to analyze future energy and demand scenarios for the sectors of the islands, three scenarios can be simulated for each sector.

Figure 1. Schematic Diagram of ISLA Model

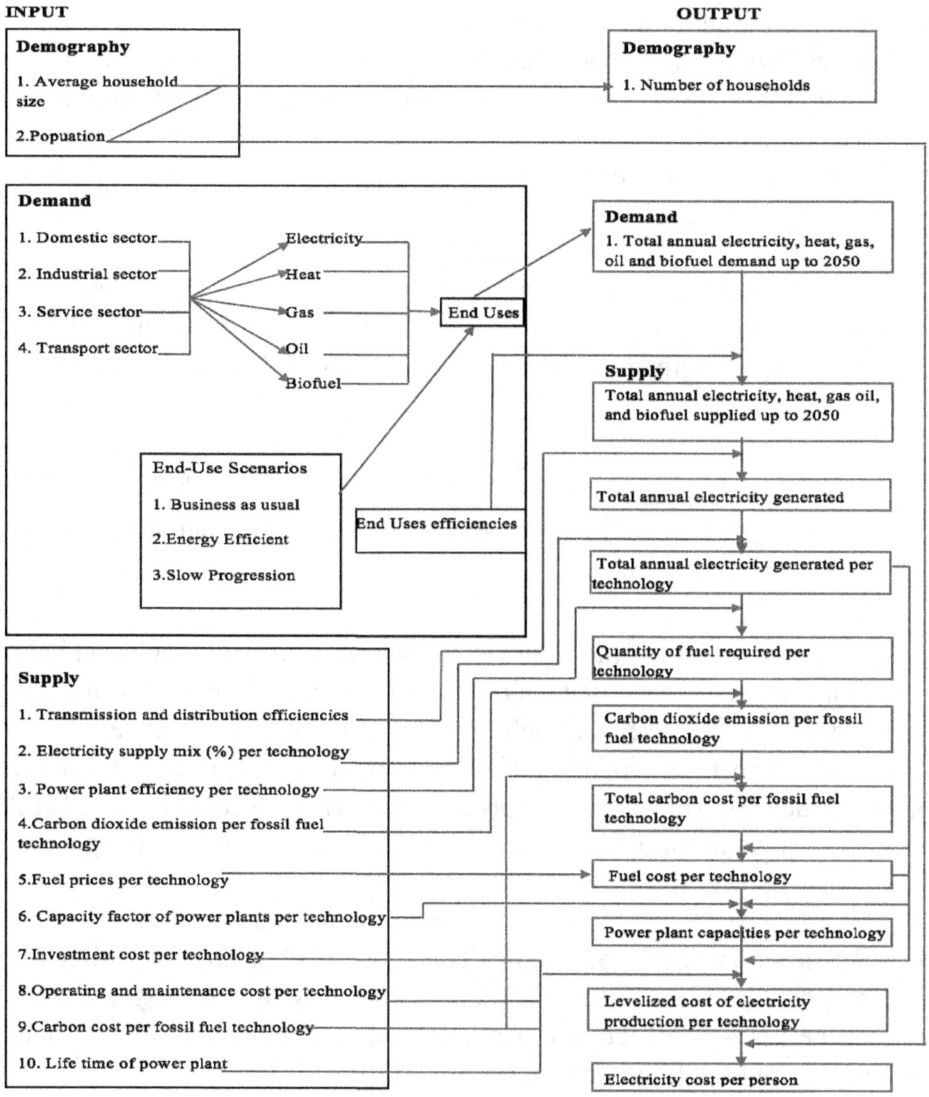

INPUT
Demography
1. Average household size
2.Popuation
Demand
1. Domestic sector
2. Industrial sector
3. Service sector
4. Transport sector
Electricity
Heat
Gas
Oil
Biofuel
End Uses
End-Use Scenarios
1. Business as usual
2.Energy Efficient
3.Slow Progression
End Uses efficiencies
Supply
1. Transmission and distribution efficiencies
2. Electricity supply mix (%) per technology
3. Power plant efficiency per technology
4.Carbon dioxide emission per fossil fuel technology
5.Fuel prices per technology
6. Capacity factor of power plants per technology
7.Investment cost per technology
8.Operating and maintenance cost per technology
9.Carbon cost per fossil fuel technology
10. Life time of power plant
OUTPUT
Demography
1. Number of households
Demand
1. Total annual electricity, heat, gas, oil and biofuel demand up to 2050
Supply
Total annual electricity, heat, gas oil, and biofuel supplied up to 2050
Total annual electricity generated
Total annual electricity generated per technology
Quantity of fuel required per technology
Carbon dioxide emission per fossil fuel technology
Total carbon cost per fossil fuel technology
Fuel cost per technology
Power plant capacities per technology
Levelized cost of electricity production per technology
Electricity cost per person

Simulation Scenarios

- **Business as Usual Scenario:** This is a scenario where historical trends from up to 10 years continues into the future.
- **Energy Efficient Scenario:** This s a scenario where there are policies and incentives in place to reduce energy demand and increase renewable energy generation. It is assumed that energy consumers are more conscious of their carbon footprint and are actively trying to reduce their carbon emission, by spending more money on efficient appliances and home energy management.
- **Slow Progression Scenario:** This scenario assumes low economic growth for the island and thus limited funds to implement environmental friendly and low carbon policies. Only few energy consumers can afford to be proactive in being energy efficient, by replacing old appliances with more efficient ones.

The third part of ISLA model focuses on the electricity supply of the island. Given the losses in transmission and distribution lines, the total electricity supplied to meet demand on the island is calculated using the following formula;

$$Total\,Electricity\,Supplied = \frac{Total\,Electricity\,Consumed}{\left(Transmission\,Efficiency + Distribution\,Efficiency\right)}$$

The electricity mix of the island from the different technologies and also interconnections with mainland of the country or other islands are simulated. Interconnections are useful in helping islands reduce dependence on their fossil fuel supply technologies and improves their security of supply in cases of emergency. Two electricity supply mix scenarios are considered in ISLA in order to account for polices, technology growth, investments and planned shutdown of power plants.

- **Business as Usual Scenario:** This scenario considers existing and proposed power plants for the islands while also taking into the account of the life time of the power plants to determine when a power plant will be put out of commission.

- **Renewable Energy Scenario:** This scenario focuses on policies and investments that encourage installation of more renewable technologies. The potential of renewable resources on the island and technology growth of these resources is also considered in this scenario.

The required quantity of fuel for the different electricity supply technologies and the carbon-dioxide emission from the fossil power plants are calculated using the following formulas;

$$Fuel\,Input = \frac{Electricity\,supply\,from\,power\,plant}{Power\,plant\,efficiency}$$

$$Carbon\,Dioxide\,Emission = Fuel\,Input \times Rate\,of\,Carbon\,Dixide\,Emission$$

The fourth part of ISLA focuses on the economics aspect of energy supply of the island.

Levelized cost of Energy (LCOE) is used in the model to calculate the present value of the total cost of building and operating a power plant over an assumed lifetime. LCOE is used because it is suitable for comparing different supply technologies of varying life spans, project size, capital cost, risk, return, and capacities. The formulas used in the economic section are as follows;

$$Power\,Plant\,Capacity(GW) = \frac{Annual\,Energy\,Output(TWh)}{Operating\,Hours \times Capacity\,Factor}$$

$$\begin{aligned}&Net\,present\,value\,of\,total\,cost\,of\,production(£)\\ &= \frac{Total\,investment\,cost(£) + Total\,operation\,and\,maintenance\,cost(£)}{Lifetime\,of\,power\,plant(years)}\\ &+Total\,fuel(£) + Carbon\,cost(£)\end{aligned}$$

$$LCOE(P\,/\,kWh) = \frac{Net\,present\,value\,of\,total\,cost\,of\,production(£) \times 100}{Net\,present\,value\,of\,production}$$

The fifth and final part of ISLA focuses on the economic dispatch of electricity supply in the island. In order to carry out economic dispatch, several

Box 1.

$$\textbf{Gaspowerplant dispatch} = IF(Hourlydemand - sum(Basesupply) > 0, IF(Hourlydemand) - sum(Basesupply) < Availablegassupply, (Hourlydemand) - sum(Basesupply), Availablegassupply)), 0)$$

$$\textbf{Biomasspowerplant dispatch} = IF(Hourlydemand - sum(Basesupply + Gaspowerplantdispatch) > 0, IF(Hourlydemand) - sum(Basesupply + Gaspowerplantdispatch) < Availablebiomasssupply, (Hourlydemand) - sum(Basesupply + Gaspowerplantdispatch), Availablebiomasssupply)), 0)$$

$$\textbf{Coalpowerplant dispatch} = IF(Hourlydemand - sum(Basesupply + Gaspowerplantdispatch + Biomasspowerplant\ dispatch) > 0, IF(Hourlydemand) - sum(Basesupply + Gaspowerplantdispatch + Biomasspowerplant\ dispatch) < Availablecoalsupply, (Hourlydemand) - sum(Basesupply + Gaspowerplantdispatch + Biomasspowerplant\ dispatch), Availablecoalsupply)), 0)$$

$$\textbf{Oilpowerplant dispatch} = IF(Hourlydemand - sum(Basesupply + Gaspowerplantdispatch + Biomasspowerplant\ dispatch + Coalpowerplant\ dispatch) > 0, IF(Hourlydemand) - sum(Basesupply + Gaspowerplantdispatch + Biomasspowerplant\ dispatch + Coalpowerplant\ dispatch) < Availableoilsupply, (Hourlydemand) - sum(Basesupply + Gaspowerplantdispatch + Biomasspowerplant\ dispatch + Coalpowerplant\ dispatch), Availableoilsupply)), 0)$$

constraints have to be met. The first constraint is that total electricity supply has to meet electricity demand at every hour of the year, while ensuring total supply cost is minimized. This is achieved by first dispatching electricity supply technologies in an order, starting from the least cost technology. The second constraint is that the maximum available capacities of the power plants are not exceeded.

After the base electricity demand has been met by the least cost electricity supply technologies such as nuclear, solar, wind, hydro and electricity import, the remaining supply technologies are then dispatched using the formulas shown in Box 1.

The results from ISLA are illustrated using Sankey diagrams, which shows the flow of energy in the island. Key aspects of a Sankey diagram are;

1. The width of the 'arms' represents the energy transferred, but the length of the 'arms' does not.
2. The input is from the left of the diagram and the output is to the right
3. The total input always equals the total output

The results from ISLA are generated into Sankey diagrams using a web tool on http://sankeymatic.com . The format for inputting data on this web tool are;

```
Source [input data] Target
Total input = Total Output
In order to construct the Sankey diagram I suggest to follow
the following steps
```
Step 1: `Go to http://sankeymatic.com`
```
Note
Source [VALUE] Target
Total input = Total Output
```
Step 2: `Create the flow for energy supplied in the baseline year you choose (ktoe);`
```
1.        Oil
2.        Gas
3.        Coal
4.        Nuclear
5.        Wind
6.        Electricity trade
7.        Biofuel
8.        Hydro
9.        Solar
10.       Biomass
```
Step 3: `Create the flow for each of the energy sources consumed`

by their respective end users,
Oil;
Thermal generation
Electricity
Losses
Road (passengers)
Road (freight)
Air
Rail
Water Transport
High temp process
Low temp process
Drying/ separation
Space heating (industrial)
Others (industrial)
Catering
Cooling and Ventilation
Water heating (services)
Space heating (services)
Others (services)
Step 4: Repeat the previous step for the other energy sources; Gas, Coal, Nuclear, Wind, Electricity trade, Biofuel, Hydro, Solar, and Biomass.
Step 5: Create the flow for energy consumed in the baseline year by the following sectors (ktoe);
Domestic
Industry
Services
Transport
District heating
Losses
Step 6: Create the flow for each of the sectors consumed by their respective end users, starting with the
domestic sector;
Space heating (domestic)
Water heating (domestic)
Cooking
Lighting (domestic)
Appliances
Step 7: Repeat the previous step for the other sectors; Industry, Services, Transport, District heating, Losses.
Step 8: Categorize End users by type;
Space heating: Space heating (domestic) + Space heating (industrial) + Space heating (services)
Water heating: Water heating (domestic) + Water heating (services)
Lighting: Lighting (domestic) + Lighting (Industrial) + Lighting (services)
Others: Others (industrial) + Others (services)

To test the validity of our approach, Crete island has been chosen. The case of Crete is explained in detail in the next section

APPLICATION OF ISLA MODEL: THE CASE OF CRETE

The Greek islands possess a magnificent beauty and they attract high levels of tourism every year. However, the Greek islands face also a number of challenges such as; migration, refugee crisis, extreme weather conditions, increasing level of tourists and high volume of waste). The Island of Crete is the largest island and the most populated in Greece, with a population of more than 600,000 people. Crete is a very popular island in terms of tourism. Crete is located at the southern part of Greece. It is the only island in Greece considered a small isolated system since 2014 when the Non-Interconnected Islands (NIIs) Code entered (Hellenic Republic,2014), being excluded from Article 2(27) of Directive 2009/72/EC. All other Greek NIIs are qualified as isolated micro-systems (European Commission, 2014).

Table 3 highlights some demographic and energy details about the island of Crete. Crete island is located in the Mediterranean climate zones, with temperatures rising up to 33 degrees in the summer period and as low as 13 degrees in the winter period, as shown in Figure 2.

Table 3. Crete Island Characteristics

Population	622,61 thousands
Area	16.906,302 km2
Density	65,96 persons/km2
3 autonomous power stations (APS)	capacity 819.25 MW
Wind Projects Solar Projects Hydro Projects	203 MW 78 MW 0,3 MW
Total variable cost (2012-2015)	185€/MWh
Annual Public Service Obligation	Circa 403 Million € based on 2012 latest available data
Power Generation of Crete shares	50% of the total Generation in the NII
Directives to set restrictions to max generation	2010/75/EU and 2015/2193/EU

Figure 2. Average temperature profile in Crete (2015)

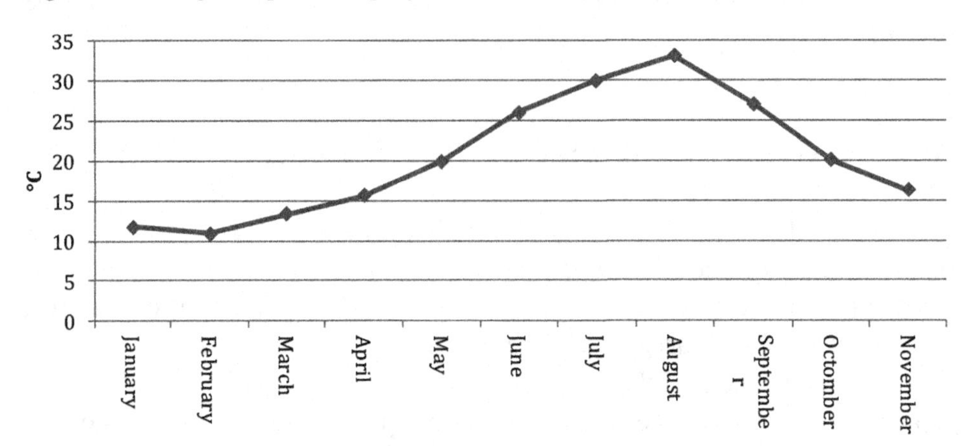

Figure 3. Electricity consumption by source in Crete Island

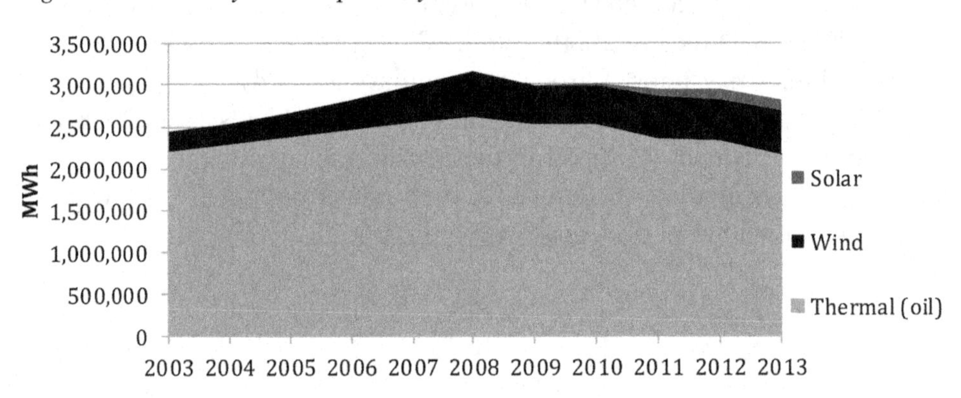

Given that the southern part of Crete is deeper in the Mediterranean, these areas experience a more constant sunny climate throughout the year, while the western part of Crete experiences more rainfall than other parts of the island.

The electricity consumption trends for a 10 years period is shown in Figure 3, with 88% generated from the oil, 5% from wind, 4% from biomass and the remaining 3% from other renewable sources. The total installed capacity is approximately 820MW. For the electricity consumed on the island, 59% is consumed in the service sector, 33% in the domestic sector, while the remaining 8% is consumed in the industry as illustrated in Figure 4. The other types of energy used on the island are oil and gas, and consuming in total 428 ktoe in 2015 on the island. From Figure 4, the transport sector consumed the highest share with 71%, followed by domestic, industry and services with 15%, 8% and 6% respectively. This also emphasizes the role of oil on the Crete Island.

All Greek non-interconnected islands could be considered isolated micro-systems except Crete. The only producer and supplier of thermal power generation is the Public Power Cooperation (PPC). The Island System Operator following the unbundling law 4001/2011 is the HEDNO. Crete operates according to a similar structure with the NGS under the provisions of the NNI Code (2014) Liberalization of the electricity system of the Non-interconnected islands enacted with Law 2273/1999 and with the Ministerial Decision n. 39/2014. Renewable energy sources projects in Crete (and other NNI) had a 25% premium above the FIT for the NGS (due to power curtailment) – cancelled with law 4416/2016.There are plans for future interconnection, First by 2020, an AC 2X200 MVA and by 2024, a DC 2X350 MW.

For the ISLA model, the baseline year considered is 2015, with a population of 622,910 and average household size of 2.8 people. As mentioned earlier, the second part of the model focuses on the energy demand of the different sectors on the island. Based on data availability, Crete has been classified into domestic, industrial, services (which includes commercial, public and agriculture), tourism and transport sectors. For the domestic sector, energy consumed is split into electricity, oil and gas. The electricity consumed the domestic sector is also spilt into its end uses which include: space heating, water, cooking, lighting, cooling, appliances and others. The oil consumed in the domestic sector is split into water and space heating, while the gas consumed is used only for cooking. For the industrial sector, energy consumed is split into electricity, oil and gas. The electricity consumed in the industrial sector is also spilt into its end uses which includes: heat processes, engines, extruders, motors, compressed air, lighting, refrigeration, space heating and others. On the other hand, oil and gas consumption were split into space heating, water and cooking only. For the service sector, energy consumed is split into electricity, oil and gas. The electricity consumed in the service sector is also spilt into its end uses which includes: catering, pumping, cooling and ventilation, water, space heating, irrigation, lighting and others. While the oil and gas end use consumption are similar to the industrial sector. For the tourism sector, energy consumed is only electricity and its end uses are: accommodation, catering, water supply production/commercial, and drainage. Finally, for the transport sector, the only type of energy consumed was from oil. For the third part of the model, the electricity supply mix is 79.5% from oil, 15.3% from wind, 4.9% from solar, 0.3% from biomass, and 0.021% from hydro.

Results With ISLA Model for Crete Island

The population for Crete island is assumed to have a yearly growth of 0.6% per year from 2016 to 2030 and decreases to 0.4% yearly from 2031- 2050. On the other hand, average household size is assumed to be constant 2.8 from present to 2030, and decreases to 2.65 from 2031- 2050. Figure 4 shows the simulated population and number of households up to 2050. Given that the population increases and the household sizes remain the same up to 2030, there is a decrease in the total number of households in Crete. As the household size decreases from 2031, there is a noticeable increase in the number of households, and by 2050 the number of households is almost back to the same level as in 2015.

Figure 5 illustrates an energy efficient scenario in the domestic sector for Crete island up to 2050. As a result of increase in population, there is significant increase in the consumption of energy for all the end uses from 2016. However, from 2031, automatic controls of lighting in homes and significant replacement of old lighting with LED, results in reduced electricity consumption for lighting. The energy efficient scenario also assumes that from 2031 more efficient appliances will be bought on the Island and more residences will make use of home management systems. Space heating also reduces as a result of better insulation in homes. Energy consumption in the domestic sector is majorly from water heating, followed by cooking and then appliances, while the least energy consumed is from lighting. This is why

Figure 4. Population and household number projected for 2015-2050

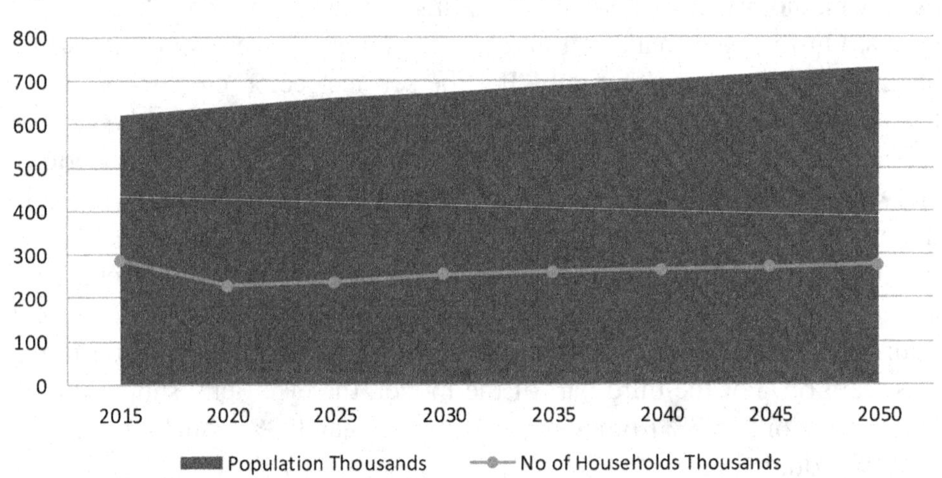

Figure 5. Domestic Energy Consumption per end use, 2015-2050

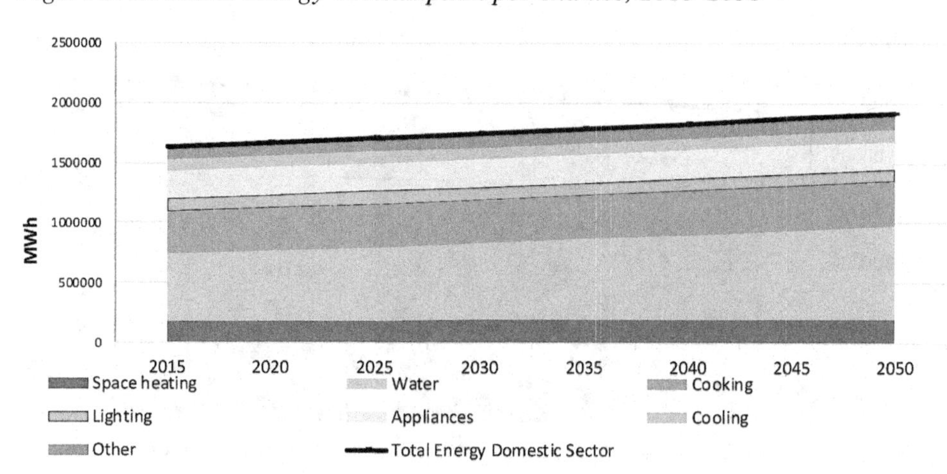

despite steps in making lighting systems more energy efficient, the impact is not very noticeable in the entire energy consumption in the domestic sector.

An energy efficient scenario in the industrial sector for Crete island up to 2050 is shown in Figure 6. Similar to the domestic sector, increase in population results in the energy consumed for certain end uses such as heat processes, extruders, compressed air, refrigeration, space heating and others. However, in an efficient scenario with better lighting controls, LED lighting, technology growth for heat pumps, engines and motors there is a slight decline in the energy consumption from these four end uses. The highest end use for energy consumption in the industrial sector is the heat process, followed by motors and then refrigeration. On the other hand, lighting and extruders consume the least amount of energy in the industrial sector. Given that motors are the second largest consumers of energy in this sector, having more energy efficient motors does have an impact on the overall energy consumption of the sector.

Similar to the previous sectors, the energy consumption for space heating and lighting in buildings and streets decreases in the energy efficient scenario for the service sector, as shown in Figure 7. Energy consumption from other end uses in the service sector is expected to increase due to increase in population and economic activities, however with more increasing production of more efficient agricultural products, energy consumption for irrigation in particular will experience a decline. In 2015, the highest consumption of energy from the service sector was from cooling and ventilation, followed closely by catering. However by 2050, given the increase in economic

Figure 6. Industrial energy consumption per end use, 2015-2050

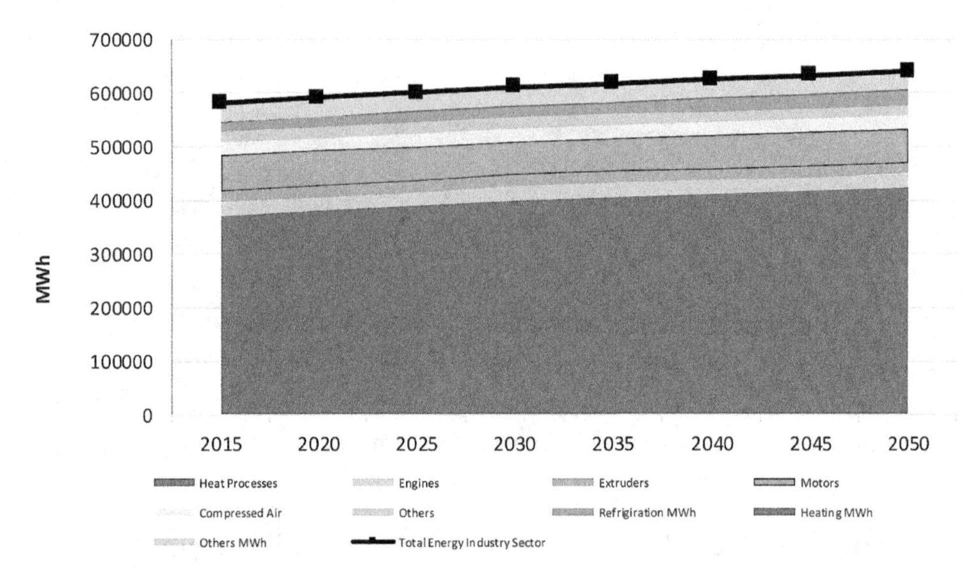

activities catering consumes the most energy. The energy efficient activities in the agricultural sector has a slight impact on the total energy consumed, therefore the focus of energy efficiency polices should be on catering and cooling and ventilation.

Figure 7. Services energy consumption per end use, 2015-2050

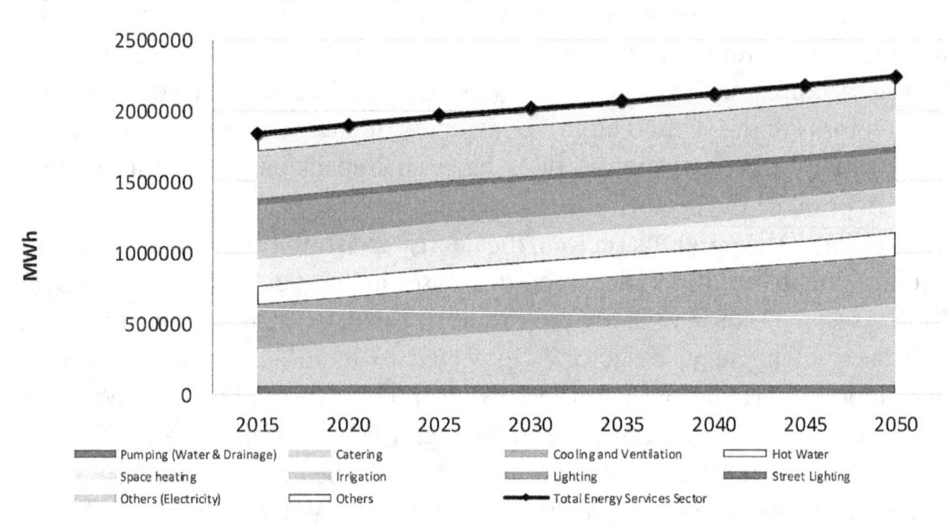

Figure 8. Tourism sector energy consumption per end use, 2015-2050

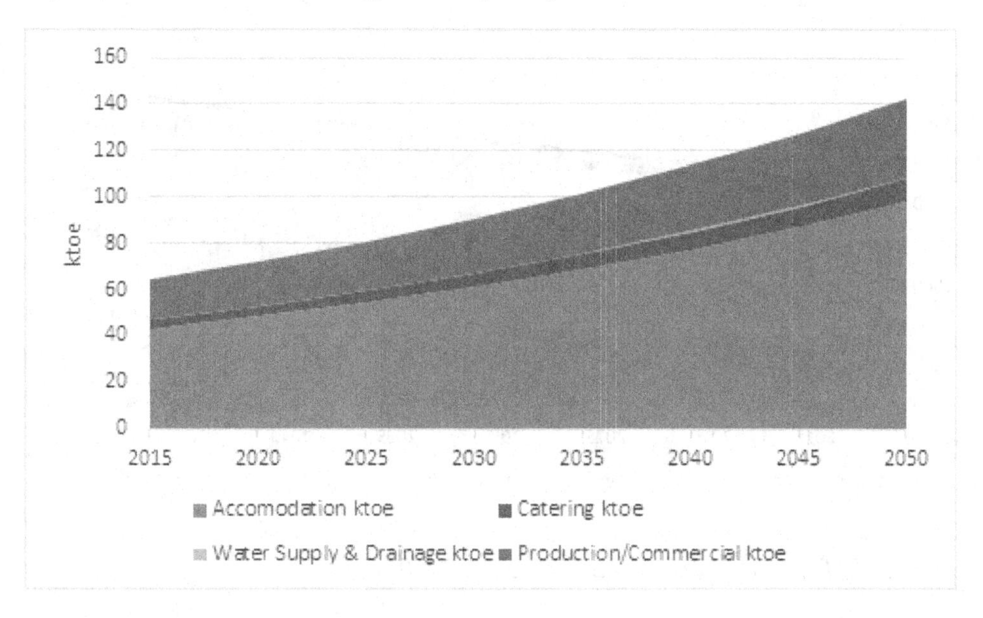

Presently, Crete is one of the most popular holiday destinations in Europe and this is expected to continue as economic growth in Europe increases annually. As tourism increases, not only do more people visit the island but it is expected that inhabitants of the islands will also experience economic growth and invest more the tourism activities. This directly impacts the production and commercial aspect of the tourism sector. The highest energy consumption is from accommodation, followed by production and finally catering (Figure 8). This trend continues from 2016 to 2050 as expected.

Figure 9 illustrates the total electricity consumed in the energy efficient scenario for all the sectors in Crete island up to 2050. Electricity consumption in the service sector is the highest, followed domestic sector, then industrial and then transport sector. The transport sector is expected to start consuming electricity by 2020 with the introduction of electric vehicles on the island. This figure shows that more energy efficient polices need to be focused on the service and domestic sector in order for there to be a significant decrease in the consumption of electricity.

The electricity supply mix for Crete island shown in Figure 10 is a renewable scenario where there is an increase in the integration of renewable energy sources based on the potential of the resources on the island. The share of wind and solar resources in the supply mix increases from 15.3% and 4.9% to 16.5% and 5.2% respectively. By 2050, hydro has a share of 0.1% in the

Figure 9. Total electricity consumption per sector, 2015-2050

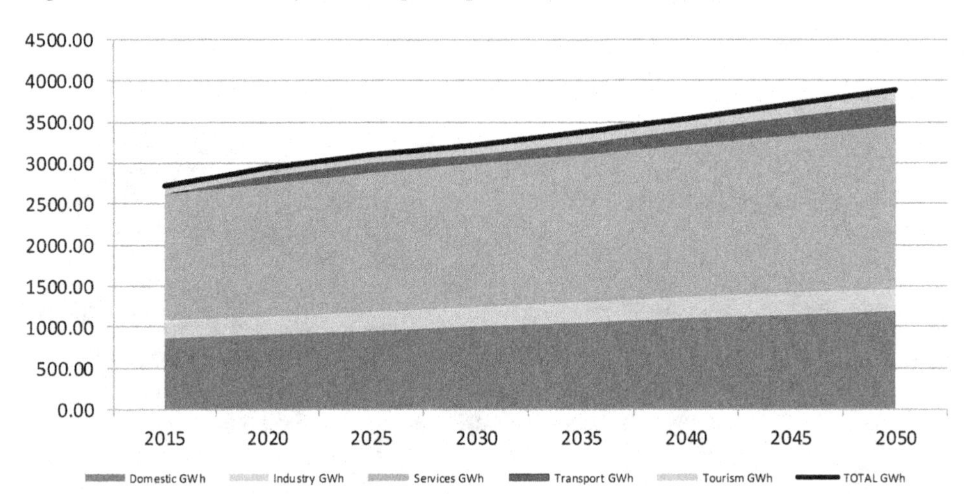

Figure 10. Electricity supply mix (%), 2015-2050

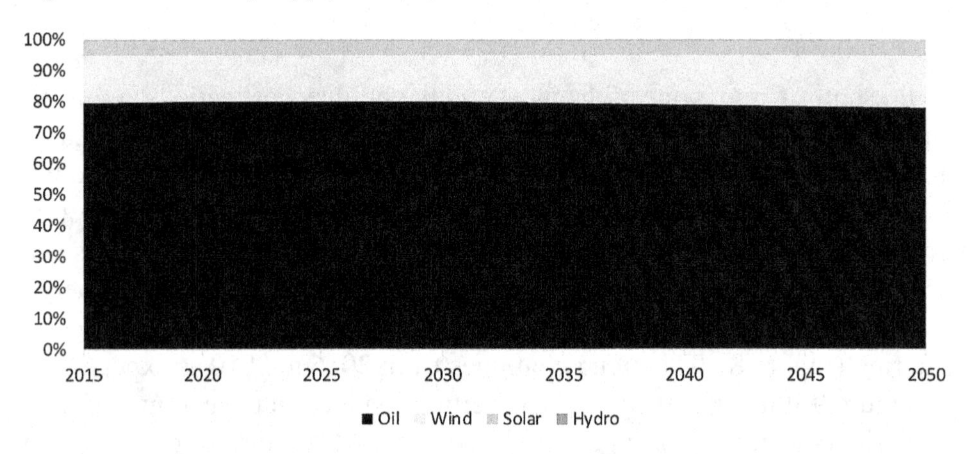

mix. The share of oil in the mix from 2015 to 2050 barely changes with a little decrease from 79.8% to 78.2%.

The results from the ISLA model is compared with results from a study by National Technical University of Athens (NTUA) on Crete Island as shown in Figure 11. Scenario 1 in the ISLA model refers to business as usual scenario based on existing power plants and proposed plans for new power plants while Scenario 2 refers to the renewable scenario where there is an increase in renewable energy integration in the Crete system. The electricity generated from oil and wind power plants in 2050 from the NTUA study is 5% and 27% higher that the levels in ISLA' renewable scenario respectively. On the

Figure 11. Electricity supply mix 2015-2050: Comparison results between ISLA model and NTU model

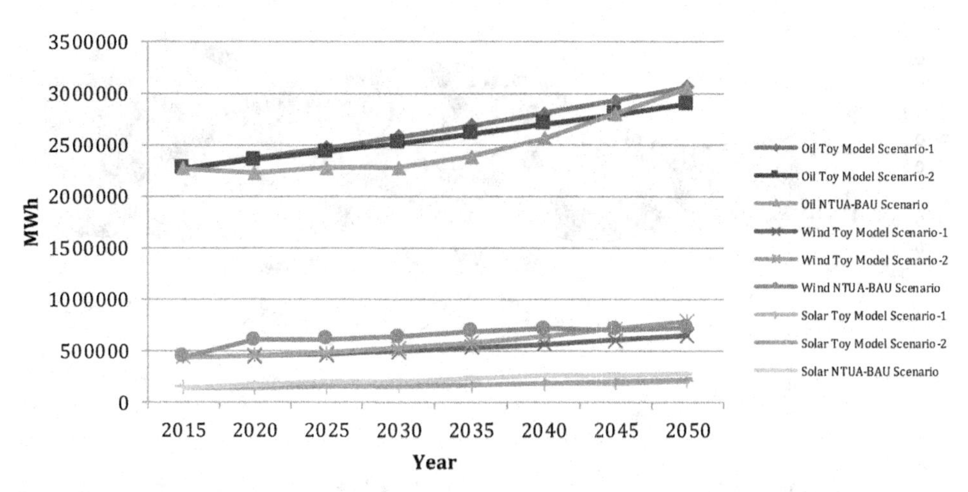

Figure 12. The levelized cost of energy

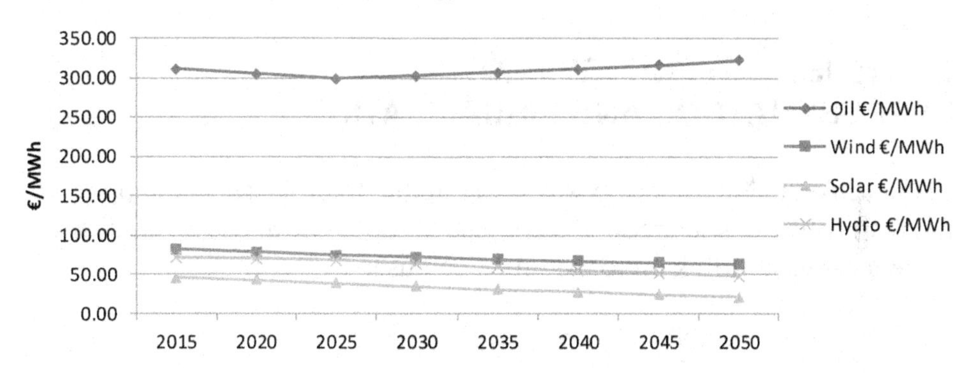

other hand, the electricity generated from wind power plants in 2050 from the NTUA study is 7% lower than the levels in ISLA's renewable scenario.

The Levelized cost of Energy (LCOE) considering different supply technologies of varying life spans, project size, capital cost, risk, return, and capacities is shown in Figure 12.

Dispatch of the oil generation is subject to the two constraints. The first constraint is that total electricity generation meets demand at every hour and the second constraint is that maximum available capacities of generation are not exceeded. Dispatch of the power plants to meet demand at every hour is carried out while ensuring total generation cost is minimized. An example

Figure 13. Generation dispatch for one day 24th April 2014

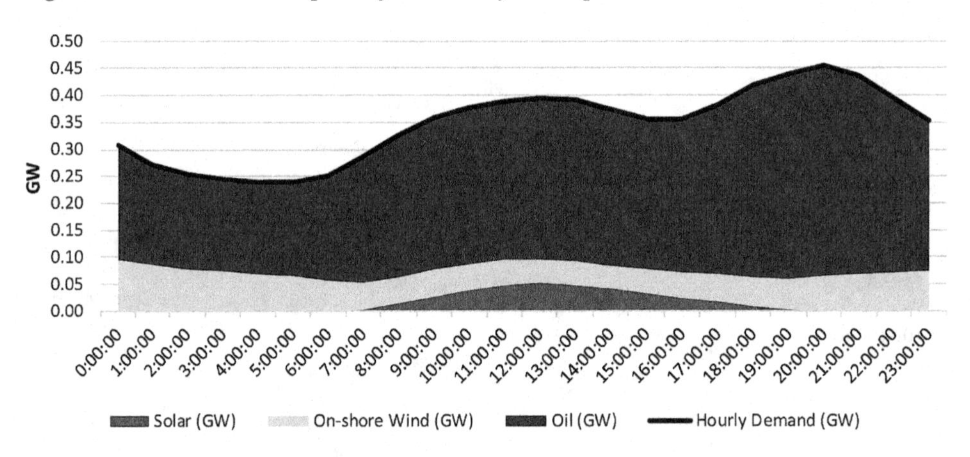

of generation dispatch model for one day (24th April 2014) is provided in Figure 12.

CRETE ISLAND ENERGY FLOW REPRESENTATION AND VISUALISATION

Using sankeymatic, first step was to create the flow for energy supplied in 2015 from various sources. Exceptionally, since no disaggregation takes place for transport link oil to the direct end user

```
Oil [341.87] Thermal Generation
Oil [305.25] Road Transport
Gas [8] Thermal Generation
Wind [39.86] Electricity
Solar [12.75] Electricity
Hydro[0.06] Electricity
Biomass [0.78] Heating
```

Second step was to create the flow for each of the energy uses consumed by their respective end users

```
Thermal Generation [207.11] Electricity
Thermal Generation [142.76] Heating
Electricity [89.76] Domestic
Heating [81.43] Domestic
Electricity [22.31] Industry
Heating [34.97] Industry
```

```
Electricity [122.43] Services
Heating [27.45] Services
Electricity [25.98] Transmission Losses
```

Next step step was to create the flow for energy consumed in 2015 by the Domestic sector

```
Domestic [8.23] Space heating
Domestic [25.59] Water
Domestic [23.92] Cooking
Domestic [9.5] Lighting
Domestic [19.76] Appliances
Domestic [18.36]Losses
Domestic [65.83] Other heating
```

Then repeat the previous step for the other sectors; industry, services

```
Industry [31.82] High Temp Process
Industry [4.01] Low Temp Process
Industry [0.71] Drying/Separation
Industry [5.84] Motors
Industry[2.04] Compressed Air
Industry[1.71] Lighting
Industry[1.43] Refrigiration
Industry[0.19]Space heating
Industry[2.92] Others
Industry [6.6] Losses
Services [21.89] Catering
Services [13.53] Pumping (Water, Drainage) & Irrigation
Services [22.57] Cooling and Ventilation
Services [21.49] Hot Water
Services [22.57] Space heating
Services [25.61] Lighting
Services [9.41]Others
Services [12.82] Losses
```

We then categorize end users by type

```
Space heating: Space heating (domestic) + Space heating
(industrial) + Space heating (services)
Water heating: Water heating (domestic) + Water heating
(services)
Lighting: Lighting (domestic) + Lighting (Industrial) +
Lighting (services)
Others: Others (industrial) + Others (services)
```

Figure 14. Crete Island Energy Flow Representation

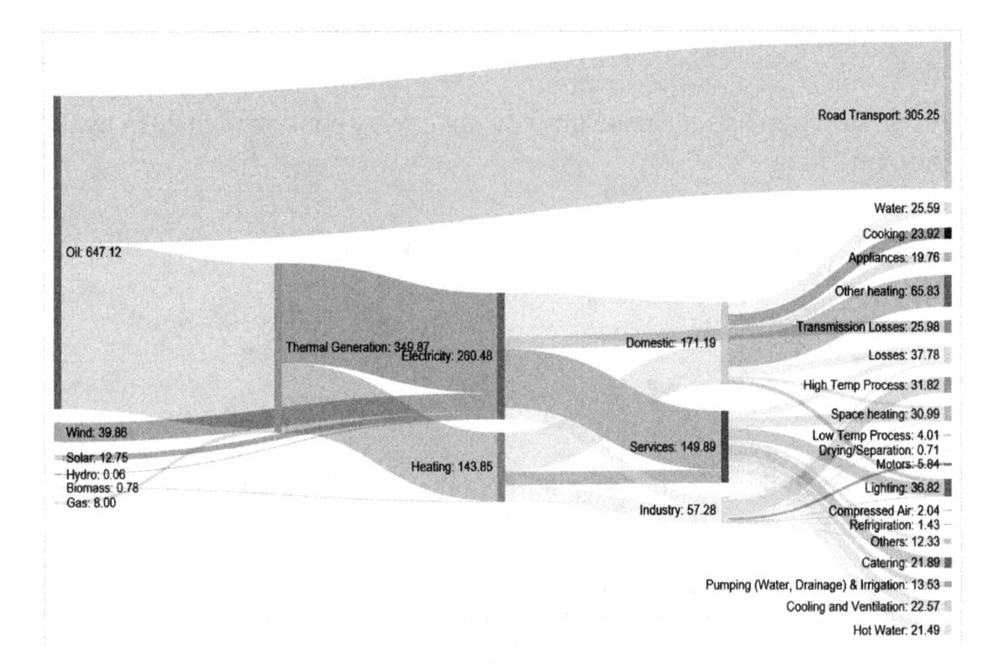

FUTURE RESEARCH DIRECTIONS

There are several things which can be improved in terms of future research directions in whole energy system modeling of islands. Models of energy systems or components have played a key role in formulating energy policy, especially as we move from developing strategies to implementing.

Whole system models enable the full representation of demand side and supply side changes to be tested and integrated across sectors, so that the impact of the interaction of policies in different areas on each other can be fully tested. In the domain of energy system modeling there is a need for a better integration between physically based models of the energy system and econometric based models.

Perhaps more important than epistemological hierarchy is the fact that many interesting problems are mostly viewed as primarily physical issues with economic consequences that can be worked through as an addendum to the physical analysis. Their key features can be expressed more quickly, powerfully and elegantly in the language of physics and engineering than in the language of economics.

CONCLUSION

There are philosophical underpinnings to models in terms of questions, methods and coverage. Models need interesting questions and applications defined first and development second. What is hard is to define practicable modeling and data projects that usefully answer questions with the requisite experts. ISLA toy model proved that can provide results as accurate as advanced models. The central aim was to develop a workable model to help design and understand the future energy system evolution for any island in the world. The toy model has been proved to be a great vehicle for teaching, and in the same time ensuring better transfer of knowledge about the models to students and those who are interested in learning. Modelling whole energy systems, however, is a great challenge mainly due to the inherent uncertainty present in the systems. Energy demand is driven not just by the conscious decisions, but also by ingrained behavioural patterns of consumers. If we add in the element of technology, and the process of energy systems modeling becomes increasingly complex.

REFERENCES

Bakos, G. C., & Soursos, M. (2002). Technical Feasibility and Economic Viability of a Grid-connected PV Installation for Low Cost Electricity Production. *Energy and Buildings, 34*(7), 753-58. Retrieved from http://www.sciencedirect.com/science/article/pii/S0378778801001426

Carta, J. A., Gonzalez, J., & Subiela, V. (2003). Operational analysis of an innovative wind powered reverse osmosis system installed in the Canary Islands. *Solar Energy, 75*(2), 153–168. doi:10.1016/S0038-092X(03)00247-0

Centre for Energy. (n.d.). *Environmental and Systems Analysis, ENPEP Applications in Asia*. Retrieved from: http://ceeesa.es.anl.gov/news/EnpepwinAppsAsia.html

Chen, F., Duic, N., Alves, L. M., & Carvalho, M. (2007). Renewislands-Renewable Energy Solutions for Islands. *Renewable and Sustainable Energy Reviews, 11*(8), 1888-1902. Retrieved from http://www.sciencedirect.com/science/article/pii/S1364032106000232

Connolly, D., Lund, H., Mathiesen, B. V., & Leahy, M. (2011). The First Step Towards a 100% Renewable Energy-System for Ireland. *Applied Energy, 88*(2), 502-07. Retrieved from http://www.sciencedirect.com/science/article/pii/S030626191000070X

Das, A., & Ahlgren, E. (2007). *Analysis of the Impact of Enhanced Use of Renewable and Advanced Fossil Fuel Technologies for Power Generation in Selected ASEAN Countries and Development of Appropriate Policies and Institutional Frameworks.* Tech. EC-ASEAN Energy Facility. Retrieved from: https://iea-etsap.org/docs/Chalmers-EC-Asean-2007-summary.pdf

Demiroren, A., & Yilmaz, U. (2010). Analysis of Change in Electric Energy Cost with Using Renewable Energy Sources in Gökceada, Turkey: An Island Example. *Renewable and Sustainable Energy Reviews, 14*(1), 323-33. Retrieved from: http://www.sciencedirect.com/science/article/pii/S1364032109001348

European Commission, *Commission Decision of 14 August 2014 granting the Hellenic Republic a derogation from certain provisions of Directive 2009/72/EC of the European Parliament and of the Council. Brussels: Official Journal of the European Union*, 2014, pp. 12-27

Giatrakos, G. P., Theocharis, D. T., & Zografakis, N. (2009). *Sustainable Power Planning for the Island of Crete. Energy Policy, 37(4),* 1222–1238. Retrieved from http://www.sciencedirect.com/science/article/pii/S0301421508006551

Giatrakos, G. P., Tsoutsos, T., & Zografakis, N. (2009). Sustainable power planning for the island of Crete. *Energy Policy, 37*(4), 1222–1238. doi:10.1016/j.enpol.2008.10.055

Hellenic Republic. (2014). *Ministerial Decision N. 313/2014, FEK 1836.* Issue B.

International Atomic Energy Agency. (2008). *Cuba: A Country Profile on Sustainable Energy Development.* Rep. International Atomic Energy Agency, 2008. Retrieved from: http://www-pub.iaea.org/MTCD/publications/PDF/Pub1328_web.pdf

Kalogirou, S. A. (2001). Use of TRNSYS for Modelling and Simulation of a Hybrid Pv–thermal Solar System for Cyprus. *Renewable Energy, 23*(2), 247-260. Retrieved from http://www.sciencedirect.com/science/article/pii/S0960148100001762

Katsaprakakis, D. A., Christakis, D. G., Pavlopoylos, K., Stamataki, S., Dimitrelou, I., Stefanakis, I., & Spanos, P. (2012). Introduction of a wind powered pumped storage system in the isolated insular power system of Karpathos-Kasos. Applied Energy, 97, 38-48.

Koroneos, C., Michailidis, M., & Moussiopoulos, N. (2004). Multi-objective optimization in energy systems: The case study of Lesvos Island, Greece. *Renewable & Sustainable Energy Reviews*, *8*(1), 91–100. doi:10.1016/j.rser.2003.08.001

Leaver, J. D., Gillingham, K., & Baglino, A. (2012). *System Dynamics Modeling of Pathways to a Hydrogen Economy in New Zealand: Final Report*. Rep. Unitec ePress. Retrieved from: http://www.unitec.ac.nz/epress/wp-content/uploads/2013/01/System-Dymanics-Report-final-updated-31.1.13.pdf

Lehmann, H. (2003). *Energy Rich Japan*. Rep. Greenpeace International. Institute for Sustainable Solutions and Innovations (ISUSI). Retrieved from: http://www.energyrichjapan.info/pdf/EnergyRichJapan_summary.pdf

Mebom, P., Barth, R., Brand, H., Hasche, B., Swider, D., Ravn, H., & Weber, C. (2007). *All Island Grid Study*. Report for The Dept. of Enterprise, Trade and Investment, and the Dept. of Communications, Energy and Natural Resources. Retrieved from: http://www.uwig.org/irish_all_island_grid_study/workstream_2b.pdf

Sheldon, P. J. (2005). The Challenges to Sustainability in Island Tourism. *Occasional Paper, 2005*(October), 1. Retrieved from: http://citeseerx.ist.psu.edu/viewdoc/download?doi=10.1.1.502.5607&rep=rep1&type=pdf

Skarlis, S., Kondili, E., & Kaldellis, J. K. (2012). Small-scale biodiesel production economics: A case study focus on Crete Island. *Journal of Cleaner Production*, *20*(1), 20–26. doi:10.1016/j.jclepro.2011.08.011

Timilsina, G. R., & Shah, K. U. (2016). Filling the gaps: Policy supports and interventions for scaling up renewable energy development in Small Developing States. *Energy Policy*, *98*, 653–662. doi:10.1016/j.enpol.2016.02.028

Tsioliaridou, E., Bakos, G. C., & Stadler, M. (2006). *A New Energy Planning Methodology for the Penetration of Renewable Energy Technologies in Electricity Sector- Application for the Island of Crete. Energy Policy, 34(18),* 3757–3764. Retrieved from http://www.sciencedirect.com/science/article/pii/S0301421505002259

ADDITIONAL READING

Spataru, C. (2017). *Whole Energy System Dynamics: Theory, modelling and policy*. Routledge. Retrieved from https://www.amazon.co.uk/Whole-Energy-System-Dynamics-modelling/dp/1138799904/ref=sr_1_1?s=books&ie=UTF8&qid=1525443418&sr=1-1&refinements=p_27%3ACatalina+Spataru

Chapter 6
Islands:
Sustainable Hubs of the Future

ABSTRACT

In this chapter, the author explores favorable features of the "hub" concept for islands, including the ability to exploit the meaning of "circular energy resource hubs" approach, where the term "hub" as indicated in literature means the use of multiple energy carriers as interfaces between energy producers, consumers, and transport infrastructure. The author expands the idea to "circular energy resource hubs" by integrating circular economy principles in which we keep resources in use for as long as possible, extract the maximum value from them while in use, then recover and regenerate products and materials at the end of each service life. It demonstrates some interesting advantages in terms of applicability of the concept offered by new theoretical approach. Furthermore, circular energy resource hubs could serve as interfaces between infrastructures and network participants (producers, consumers) or between different infrastructures, representing a generalization or extension of a network node.

PLURALISTIC CHARACTERISTICS OF ISLANDS AND TRANSFORMATION

Islands have been testing with modern shapes of sustainable living for a long time, by including innovative governance schemes to be more socially comprehensive, teaching islanders about key sustainability issues. Islands

DOI: 10.4018/978-1-5225-6002-9.ch006

have delivered best practices on smart and sustainable local development, providing inspiration to other insular and mainland regions globally. The living labs can offer important lessons for different geographies including rural areas, mountainous areas, towns on energy, transport, circular economy, governance and so on. However, islands themselves are confronted with diverse challenges as a result of varities in their measure, distance from the mainland, political devolution, lack of regulations and competition.

Because of the isolated nature of islands, the most practical and cost-effective energy future in these remote areas is renewable energy and support for new business models for renewable energy communities. In addition to renewable energy, storage can also significantly increase the revenue and reduce the risk of forecast uncertainty, allowing the role of backup. With careful planning and management, policy support, financial investments, islands could play the role of clean power hubs of the future.

From a system point of view, by combining and coupling different energy carriers in energy hubs keeps a number of potential advantages over conventional, decoupled energy supply. These include increase reliability, load flexibility and have positive synergies effects. Different energy carriers can be combined to provide the most cost efficient and reliable system for a particular type of island. The whole energy system can be considered as a system of interconnected energy hubs. This will help to transform islands into smart and thriving economies. Furthermore, energy hubs can serve as interfaces between energy infrastructures and network participants (producers, consumers), or between different energy infrastructures, representing a generalization or extension of a network node in an electrical system. Furthermore, extending the concept of energy hub to circular energy resource hubs, where the term "hub" indicates the use of multiple energy carriers, as interfaces between energy producers, consumers and the transportation infrastructure and "circular" indicates the use of the circular economy principles in which we keep resources in use for as long as possible, and while in use extract the maximum value, then at the end of each service life we recover and regenerate products and materials.

The next section discusses the energy hub concept and how it was used in various studies.

ENERGY HUB CONCEPT

The pace of change occurring within the energy system means that efficient resource utilization is imperative. Distributed energy systems (DES) can be a key enabling factor in meeting future energy needs. However, the introduction of DES also results in more complex requirements for design and operation. Integration of DES in existing electrical grids is not straightforward; a key challenge is how to integrate a large share of renewables, which are inherently intermittent at distribution level. Moving towards a low carbon economy, increase use of renewables, and increase use of smart information communication technologies (ICT), energy hubs could play a key role in delivering real-time reaction.

An energy hub is considered a unit where multiple energy carriers play the interface role between different energy infrastructures and loads (Geild & Andersson, 2007). Renewable energy may be generated locally (e.g. from PV) or by centralized means (a geothermal plant or a combined heat and power (CHP) plant located within the island that may be fuelled by biofuel or hydrogen). The size hub can vary, from a building, to local community, an island, a city, a country, a region. The concept is applicable to all types of energy flow, from heating and cooling to electricity, biogas and hydrogen, and may connect not only households but also (electrical) cars, commercial buildings or industry.

The inputs and outputs of an energy hub can be mathematically connected in a form of a coupling matrix (Geild & Andersson, 2007):

$$\begin{pmatrix} p_1^{out} \\ . \\ p_m^{out} \end{pmatrix} = \begin{pmatrix} c_{11} & \cdots & c_{1n} \\ \vdots & \ddots & \vdots \\ c_{m1} & \cdots & c_{mn} \end{pmatrix} \begin{pmatrix} p_1^{in} \\ . \\ p_n^{in} \end{pmatrix}$$

where

$p_1^{in},\ldots,p_m^{in}$ *and* $p_1^{out},\ldots,p_n^{out}$ are the input and output carriers

c_{ij} is the coupling coefficient between input energy *i* and output energy *j*

The expression between in- and output is

$$P_2 = C_{12}P_1$$

According to various sources, the energy hub idea was first developed in the Vision of Future Energy Networks (VoFEN) project (VoFEN project). The literature provides several publications on energy hubs operation and management, planning, evaluation and optimization of decentralized multi-energy systems where multiple energy carriers are converted, stored and dissipated (Andersson et al., 2007; Favre-Perrod et al., 2005; Anders & Vaccaro, 2011; Krausse et al., 2011). The majority of these studies focus on mathematical modeling of energy hubs, optimization and operating scheduling, providing an optimal choice of components, connection and combinations. Very often the optimization problem consists in minimizing the total energy cost in the system, within a deterministic framework of load demands, prices, efficiencies and constraints (Carradore & Bignucolo, 2008; Schulze et al., 2008). The energy hub concept helps simplify multiple system optimization (Geidl & Andersson, 2007). Financial analysis of energy hub with demand response of heat load management is evaluated in Kienzle et al. (2011) using a Monte Carlo electricity market price simulation approach. Combining energy hub with demand response could help improve reliability of the system and reduce total operation costs. However, future research is needed to understand the impact under variable conditions (price, weather, load). More price and demand scenarios need to be analyzed. Environmental costs should be factored in.

The main conclusion that can be drawn from all these studies is that the energy hub concept could provide new insights into multiple energy carrier systems, test different combinations and their flexibility to help potential for various system improvements. Furthermore the integrated energy system can be studied by defining the nodes of such a system as energy hubs. This interconnected energy hub approach would consist of various energy conversion and storage technologies (Hemmes et al., 2007) playing the role as an interface between energy demand and supply taking advantage of the technical and economic advantage of each energy carrier. Such an interconnected hubs system is shown in Figure 1.

The concept could be further refined and elaborated using realistic examples and case studies and islands could be the best candidates.

Potential solutions for islands to become energy hubs which can be analysed and tested include:

- Reduce fossil fuel use and imports and increase use of local renewable energy production and of energy from waste and new business models to support islanders to become prosumers

Figure 1. Representation of interconnected energy and resource hubs

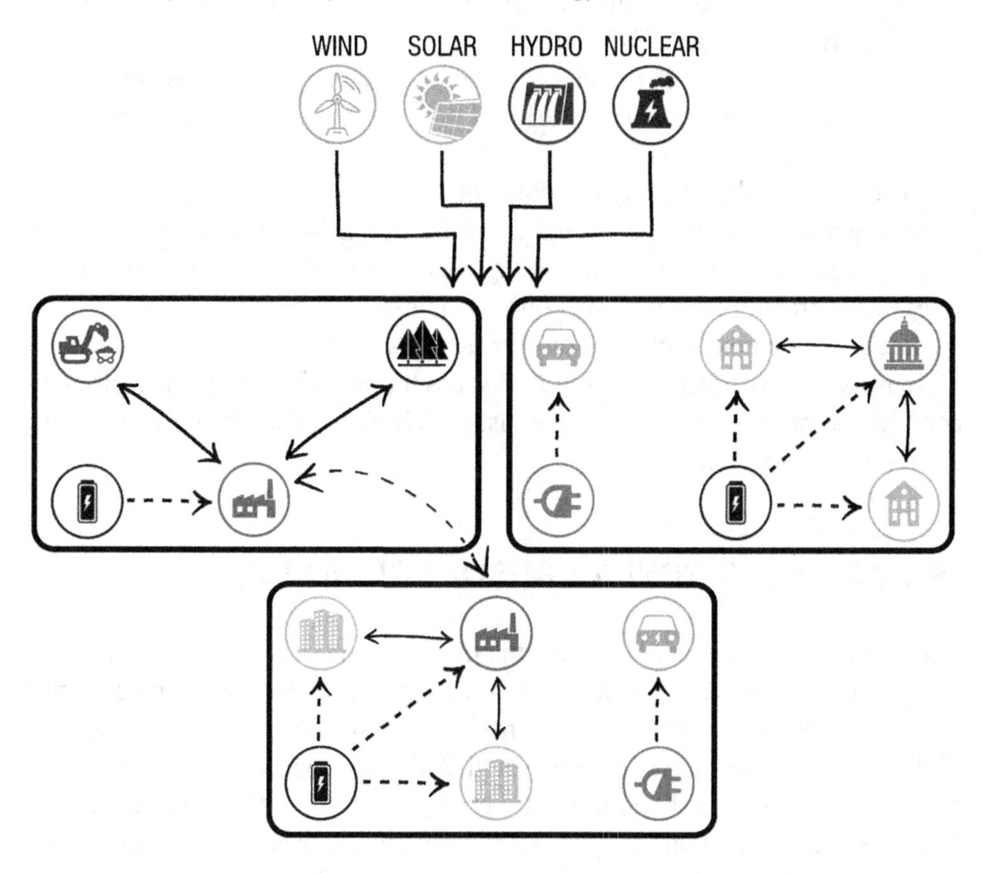

- Increase energy efficiency in infrastructures (for example street lighting and pumping stations) and of the building stock (electricity, heating and cooling)
- Realize existing synergies and connections between transport and energy, by promoting ferries using alternative fuels such as hydrogen and balancing intermittent power from renewable energy, integrating electric vehicles and ferries into islands' smart electric grids
- Change of the transport modes including car-sharing, car-pooling, promoting walking and cycling
- Pursue the transition towards zero-waste territories by adopting a circular economy development model through the strengthening of local value chains.
- Use of grey water recycling and rain water harvesting, reduce water losses

- Collect, sort, reuse and recycle waste and improve environmental quality and create jobs locally
- Introduce incentives for waste producers, in order to increase recycling rates and reduce mixed waste.

Furthermore, the energy hub concept can be expanded with alternative energy generation (e.g. bioenergy, biogas from waste) to capture the interlinkages between resources (energy, water, food, land and materials) and integrate with circular economy concept. These possible future concepts are discussed in the next section *Future research directions*.

In meantime the next section will discuss the benefits will bring a circular economy approach for islands if we installed high renewable energy and low carbon technologies.

ISLANDS AS 'CIRCULAR RESOURCE HUBS'

Circular economy concept has become more common nowadays and islands are great systems in order to accelerate self-sufficient circular economy practice. Each year, islands have been visited by millions of tourists because of their location and heritage. However, they also face various challenges such as limited resource and economic diversity. Therefore, these aspects make islands great isolated laboratories to examine circular economy principles.

Vlieland, the smallest populated island in the Netherlands, mainly rely on the mainland in terms of products and waste (Metabolic, 2017). Transportation for these streams and people has been provided by diesel-powered ferries from the island. Although Vlieland has 2020 sustainability targets, in 2014, Lab Vlieland expressed that focusing on energy self-sufficiency will not be enough to meet the targets. Therefore, Lab Vlieland and local municipality has started to collaborate in order to analyse energy, water and material flows on the island. This holistic approach could help to understand material and energy flow analysis and problems related to sustainable tourism.

Aruba, Bonaire and Curacao, which are three of the Caribbean islands, have 2025 visions for Caribbean Waste Collective (Bruinsma, 2017). The aim is to optimise the waste stream through reducing and recycling waste with the help of collaboration between stakeholders and the government. There are also several projects for future developments such as reducing food waste, establishing a central e-market platform for waste, tax facilities for waste

Table 1. Carbon footprint comparison per person per year (in tonnes)

	Samsø	Danish Average	Australian Average
Carbon Footprint	-12.0	6.2	17.0

Data source Nadine Galle, 2018.

recycling etc. This progress could help to achieve circular waste stream in Caribbean islands by enhancing employment and better use of local waste.

Several Greek islands have also good practices on circular economy in waste stream. In Tinos island, for instance, the municipality has started a project for 5 different streams (Packaging, paper/paperboard, glass, plastic & metal and biowaste) in order to enhance recycling (Loizidou 2016). Another project named PAVEthe WAySTE has been launched in Cyclades Islands (Donousa, Schinoussa, Irakleia, Koufonissi and Ancient Olympia). The aim of the project is to assist the progress of Waste Framework Directive in remote areas through increasing recycling rate (Anon, 2015). Better separation and storage techniques in micro-macro levels could help to reach better municipal waste management results in the islands.

Canary Islands have 2025 targets in terms of energy and water consumption. Current renewable share in electricity production is only 9% and the aim is to reach 45% until 2025 (Canary Islands Government, 2018). The major part of this share will be produced by solar and wind plants and supported with other renewable energies such as biogas, hydro and geothermal. The water-energy nexus and integration of the storage systems in energy production could also help to reduce fossil share and increase the resource efficiency. On the other hand, there are also several successful circular economy implications in islands. Samsø, an island in Denmark, reached its renewable energy target in 2009 (Nadine Galle, 2018). Resident-owned wind turbines help to reduce carbon footprint average to negative 12 tonnes per person per year which is lower than Danish and Australian national average (Table 1).

Moreover, Samsø has initiated "Full Circle Project" which intends to be the first fully circular place in the world. Biogas is the key motivation for 2030 targets (Tybirk et al., 2016). A biogas plant with a capacity of 100,000 tonnes/year biomass have been projected; therefore, a circular input-output chain can be established between farmlands, factories and biogas plant. Additionally, biogas plant can also cover the capacity of existing LNG-powered ferry to Jutland. A micro-plant could liquefy the biogas for ferry usage. Compressed Bio-Natural Gas can be used for trucks, busses and tractors when the vehicles

Figure 2. Circular islands approach

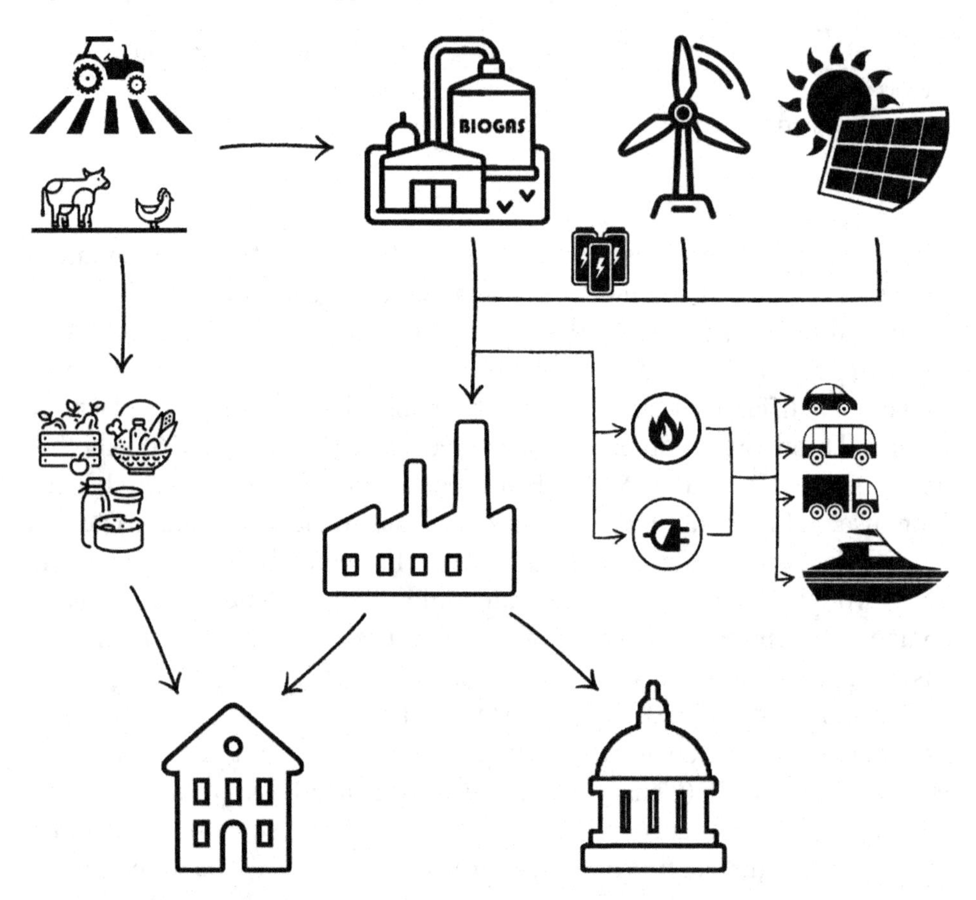

renovated for the system. As a conclusion, Biogas could be the solution of Fully Circular Samsø with the necessary commitments.

All these projects indicate that circular economy implications have considerable results on islands socially, economically and naturally (Nadine Galle, 2018). As most islands are isolated places from the mainland circular chains in material, energy and waste streams could help to reach circular economy targets easier than cities in the mainland.

A representation of a potential circular approach for islands is provided in Figure 2.

Within these an island could be formed of one hub or several hubs, depending on the various characteristics.

FUTURE RESEARCH DIRECTIONS

We argue that most islands around the world are at various stages along an evolutionary energy efficiency, transport and resources development processes which are characterized by different views about energy efficiency ownership and use, the role of different energy technologies and transport modes (i.e. due to different prevailing transport policy paradigms). This leads to the promotion of different policy objectives and measures at different points in time, contributing substantially to differences in behavior.

As mentioned above the concept of energy hub can be further expanded to other resources to capture the interlinkages between resources and integrate it with circular economy concept. We could call it "*circular energy resource hub*". The unique opportunity is the integration of multi-vector energy systems and resources, with diversity of supply which could help in increasing reliability and load flexibility. Depending on the size of the island, multiple hubs could be interconnected.

CONCLUSION

This chapter provides an understanding on the energy hubs concepts for academic research. Extracting the kind of knowledge we discussed in the present study reveals possibilities in adopting innovative approaches for islands to better understand the interplay between different resources and systems. This approach seems to outperform that of more traditional models. This is due to the power of the method to allow for non-linear and asymmetric effects. Although the method approach appears simple as representation but analytically complex, it may yield parsimonious results. Integrating other resources (water, food, materials, land use) within the energy hub concept and the circular economy principles, illustrates and confirms islands planners and project managers should identify the suitable methods to provide them the 'optimal' performance level in terms of systems choices.

Islands require integrated solutions to address these challenges while providing the infrastructures and services to enhance human living conditions and wellbeing. Enabled by the smart technologies, the combined development of renewable energy and circular economy engineering approach offers the opportunity to enhance an insular circular economy. Sources for islands energy includes renewable energy and energy recovered from municipal solid waste.

Further work is needed to identify the potential of different solutions, explore the synergies and potential trade-off of strategies that promote the transition to a low carbon circular economy in islands. This includes understanding of the composition of the energy demand in circular system. In order to achieve this it is key to understand the current situation. ISLA model could be further developed by including a diverse range of technologies and their life cycle assessment. Together with indicators based upon energy analysis, material and energy flow analysis and evaluation of footprints based on life cycle analysis and related approaches (such as ecological and carbon footprint), could help evaluate the optimal transition pathways towards sustainable island hub management. Such analysis will help provide practical guidelines for public service professionals, system planners, managers and so on.

REFERENCES

Anders, G., & Vaccaro, A. (2011). Innovations in power systems reliability. Springer. doi:10.1007/978-0-85729-088-5

Andersson, G., Geidl, M., Hemmes, K., & Zachariah-Wolff, J. (2007). Towards multi-source multi-product energy systems, sciencedi- rect. *International Journal of Hydrogen Energy*, *32*(10-11), 1332–1338. doi:10.1016/j.ijhydene.2006.10.013

Anon. (2015). *Pavethewayste*. Available at: http://www.pavethewayste.eu/en/

Bruinsma, W. H. (2017). *Turning Waste into Value - Aruba, Bonaire & Curacao leading the way, Canary Islands Government, 2018*. The Canary Islands: Natural Laboratories of Clean Energies.

Carradore, L., & Bignucolo, F. (2008). Distributed multi-generation and application of the energy hub concept in future networks. *43rd International universities power engineering conference*. 10.1109/UPEC.2008.4651593

Favre-Perrod, P., Geidl, M. B., Klockl, B., & Koeppel, G. (2005). A vision of future energy networks. *Proc. of Inaugural IEEE PES Conference and Exposition in Africa, Durban, South Africa*, 24–30.

Geidl, M., & Andersson, G. (2007). Optimal power flow of multiple energy carriers. *IEEE Transactions on Power Systems*, *22*(1), 736–745. doi:10.1109/TPWRS.2006.888988

Hemmes, K., Zachariah-Wolff, J. L., Geidl, M., & Andersson, G. (2007). Towards multi-source multi-product energy systems. *International Journal of Hydrogen Energy*, *32*(10-11), 1332–1338. doi:10.1016/j.ijhydene.2006.10.013

Kienzle, F., Ahcin, P., & Andersson, G. (2011). Valuing investment in multi energy conversion, storage and demand side management systems under uncertainty. *IEEE Trans Sust Energy*, *2*(2), 194–202. doi:10.1109/TSTE.2011.2106228

Krause, T., Andersson, G., Fröhlich, K., & Vaccaro, A. (2011). Multiple-energy carriers: Modeling of production delivery and consumption. *Proceedings of the IEEE*, *99*(1), 15–27. doi:10.1109/JPROC.2010.2083610

Loizidou, M. (2016). Good Practices of Circular Economy for Waste Management in Islands. *Circular Economy - Territorial Cohesion - Islands.* Available at: http://iswm-tinos.uest.gr/uploads/D_6_64(_)_Loizidou_presentation_14_11_2016_Heraklion_Crete.pdf

Metabolic. (2017). *Vlieland circulair*. Available at: https://www.nadinegalle.com/portfolio/circularislands

Schulze, M., Friedrich, L., & Gautschi, M. (2008). Modeling and optimization of renewables: applying the energy hub approach. IEEE international conference on sustainable energy technologies. doi:10.1109/ICSET.2008.4746977

Tybirk, K., Kristensen, U.V., Mikkelsen, N., & Stoltenborg, T.B. (2016). *From Field to Ferry – Samsø Biogas to Liquid Bio - Natural Gas for the new ferry*. Samso Energy Acadmey.

VoFEN Project. (n.d.). *Vision of future energy networks - project website.* Available from: http://www.future-energy.ethz.ch

About the Author

Catalina Spataru is an Associate Professor in Energy Systems and Networks at UCL Energy Institute in London, UK, Course Director of the MRes in Energy Demand Studies. She is the founder and the Head of the *Islands Research Laboratory*. Her research focuses on whole energy system dynamics, exploring the spatial impact of intermittent renewable energy sources, low carbon technologies, interconnections and market integration. She has led and worked on several projects on energy systems funded by the British Council, EPSRC, EC and industry (National Grid, WPD, EDF etc.) focusing on spatial and temporal energy issues. She regularly delivers presentations in academic and professional circles, public engagement events and for the media. She is the author of the book *Whole Energy System Dynamics: Theory, Modelling and Policy* published by Routledge and co-editor of Routledge *Handbook of the Resource Nexus* She teaches the following modules at UCL: *Smart Energy Systems–Theory, Practice and Implementation* (BENVGEES) (a multidisciplinary module which provides students with an understanding of the methods, concepts and practice of whole energy systems, offering a combination of theory, modelling interactive exercises and islands case studies); *Metrics, Modelling and Visualisation of the Resource Nexus module* (BENVGSR7) (a module which provides students with an understanding of modelling of the resource nexus and with techniques for output visualisation, combined with case studies for islands); *Communication Skills* (BENVGED7) (a module which provides students an introduction to academic writing, giving presentations, working with others and networking and communication with the public). She has a strong passion for art and examples of her work can be found at www.catalina-art.co.uk

Index

P

R

S

T

V

Z

Printed in the United States
By Bookmasters